Tourism—About Travelling and Vacationing in our Time

Kurt Luger

Tourism—About Travelling and Vacationing in our Time

Kurt Luger
Fachbereich
Kommunikationswissenschaft
Universität Salzburg
Salzburg, Austria

ISBN 978-3-658-44732-8   ISBN 978-3-658-44733-5 (eBook)
https://doi.org/10.1007/978-3-658-44733-5

Translation from the German language edition: "Tourismus – Über das Reisen und Urlauben in unserer Zeit" by Kurt Luger, © Der/die Herausgeber bzw. der/die Autor(en), exklusiv lizenziert durch Springer Fachmedien Wiesbaden GmbH, ein Teil von Springer Nature 2022. Published by Springer Fachmedien Wiesbaden. All Rights Reserved.

This book is a translation of the original German edition "Tourismus – Über das Reisen und Urlauben in unserer Zeit" by Kurt Luger, published by Springer Fachmedien Wiesbaden GmbH in 2022. The translation was done with the help of an artificial intelligence machine translation tool. A subsequent human revision was done primarily in terms of content, so that the book will read stylistically differently from a conventional translation. Springer Nature works continuously to further the development of tools for the production of books and on the related technologies to support the authors.

© The Editor(s) (if applicable) and The Author(s), under exclusive license to Springer Fachmedien Wiesbaden GmbH, part of Springer Nature 2025

This work is subject to copyright. All rights are solely and exclusively licensed by the Publisher, whether the whole or part of the material is concerned, specifically the rights of translation, reprinting, reuse of illustrations, recitation, broadcasting, reproduction on microfilms or in any other physical way, and transmission or information storage and retrieval, electronic adaptation, computer software, or by similar or dissimilar methodology now known or hereafter developed.
The use of general descriptive names, registered names, trademarks, service marks, etc. in this publication does not imply, even in the absence of a specific statement, that such names are exempt from the relevant protective laws and regulations and therefore free for general use.
The publisher, the authors and the editors are safe to assume that the advice and information in this book are believed to be true and accurate at the date of publication. Neither the publisher nor the authors or the editors give a warranty, expressed or implied, with respect to the material contained herein or for any errors or omissions that may have been made. The publisher remains neutral with regard to jurisdictional claims in published maps and institutional affiliations.

This Springer imprint is published by the registered company Springer Fachmedien Wiesbaden GmbH, part of Springer Nature.
The registered company address is: Abraham-Lincoln-Str. 46, 65189 Wiesbaden, Germany

Wenn Sie dieses Produkt entsorgen, geben Sie das Papier bitte zum Recycling.

Before the Journey—What can the Readers Expect?

Curiosity is the most powerful driving force in the universe, because it can overcome the two greatest braking forces in the universe: Reason and fear.
Walter Moers, *Die Stadt der Träumenden Bücher* (The City of Dreaming Books)

Authors appeal to their readership, taking them on a journey through their world of thoughts, spreading knowledge, and explaining connections. Reading as a voyage of discovery leads either into the unknown, to gain new insights, or into familiar territory, from which a special facet or a different perspective can be gleaned. An analogy could be a trip to Paris, which we may know from countless pictures—but beneath the Eiffel Tower, which looks exactly as expected, urban life unfolds with such cultural diversity that the trip remains full of new surprises, even after numerous visits.

Writing about tourism is as difficult as writing about youth: we have all been young, we have all been on the road—and we will tend to see ourselves as experts because of our individual experiences. On the other hand, this also means that basically everyone could have a say, which should make tourism a widely discussed socio-political issue. Surprisingly, this is not the case. The academic discourse on mobile leisure use has only recently emerged, and knowledge of the many interrelationships in which tourism is involved is still relatively underdeveloped.

This is all the more surprising as tourism has become a major economic sector worldwide, contributing to the prosperity of societies, and with many people dependent on its functioning smoothly. The Corona pandemic and the subsequent radical (albeit temporary) collapse of this industry have dramatically highlighted its outstanding economic importance. In 2020–2021 alone, the effects of Covid-19 on tourism resulted in a loss of over four trillion dollars to the global economy. People in the already less-advantaged countries have been hardest hit, because they cannot compensate for their income losses. In 2020 alone, there were one billion fewer international tourist arrivals, a decrease of 73%, and some 120 million jobs were lost in tourism worldwide. Indeed, tourism serves as a lifeline for millions of people. The Secretary-General of the UN World Tourism Organization WTO, Zurab Pololikashvili, hopes that through improved vaccination programmes, tourism will regain momentum within the next few years.[1] This is a hope that he shares with millions of travel and holiday enthusiasts, as well as all those who make a living from tourism.

The pandemic represents a turning point for tourism: it will probably trigger a rethink, in view of the recognized

---

[1] https://www.unwto.org/news/global-economy-could-lose-over-4-trillion-due-to-covid-19-impact-on-tourism, 27. August 2021.

Before the Journey—What can the Readers Expect?   vii

vulnerability of the system and the increasingly evident climate crisis. Tourism is more than just a matter of economics. In the reconstruction years after the Second World War, tourist travel became part of the lifestyle of Western industrial and affluent societies. Relaxation and curiosity, world experience and status gain are today the main motives of travellers, whereas generating income or profit is the interest constellation on the part of the more or less professional service providers, who still like to call themselves 'hosts'. However, tourists are *special guests*: the relationship with them is not based solely on hospitality, because the providers of such tourism services are in tough competition for every potential guest or customer. Moreover, the economic benefit also has its negative sides—for example, for those parts of the population who do not earn from tourism, but are affected by traffic, high prices in food services and housing. Unprotected nature, the environment or habitat often lose out, as tourism infrastructure consumes landscapes and drastically changes them. The aim must be a balance that includes all these aspects and relates them to each other.

In the 1980s, socio-political critique emerged against some negative impacts of tourism, particularly long-distance tourism, because of exploitation. Since then, global tourism has continued to grow: new social classes and markets, as well as individual mobility, have resulted in mass tourism. The Covid pandemic broke this trend, temporarily eliminating overtourism—but now the number of overnight stays is back to earlier levels. However, there has been resentment to tourism since its beginnings—on the part of the visited, not least because of the attitudes of the travelers—and also on the part of the travelers, reacting to the attitudes of other tourists. The critique of the 1980s was economically, sociologically, and ecologically based: the call was for *tourism with insight*. This is needed today, to counter the development of tourism, which has now taken on climate-damaging dimensions, with a long-term perspective.

What is needed is better insight into interrelationships and into our own actions, reflection on them in order to develop personal responsibility, to realize that not everything that is offered is necessarily good, but may have destructive effects. And recognition that we all, as tourists, consume resources and should perhaps reconsider our behaviour. If a flight from Vienna to Bucharest is offered for, say, a mere €7.90, that airline shows that it does not take climate change seriously and is in fact acting irresponsibly.

Insight instead of thoughtlessness in tourist behavior, and *responsibility* instead of unrestrained development can be ethical cornerstones that also make sense, economically and ecologically. Tourism is involved in climate change, not only as an affected industry, but also as a co-contributor. According to UN-WTO studies, the tourism industry accounts for around five per cent of global greenhouse gas emissions, most of which come from travel.

In affluent societies, we are accustomed to living out our contradictions largely without consequences. Driving to the airport in a petrol-guzzling SUV, flying from there to the seaport with a low-cost carrier to start a cruise, and producing overtourism at stopovers during the journey—why do we do this? Because we don't know how detrimental it is to our planet? Because we believe in our personal freedom above all else and feel that we can afford everything without being held accountable for our ecologically disastrous actions? As long as tourism traffic is powered by fossil fuels, the ecological footprint of air and cruise tourists is massively harmful to the climate. Pleasure-seeking travelers are part of the problem, co-contributors to climate change.

In an integrative view of tourism, as undertaken in this publication, global or societal phenomena must be seen and evaluated together with individual actions. How we deal with our own needs and desires and at whose expense we amuse ourselves is important, because the imperialist lifestyle of the West produces winners as well as losers and can open up vistas beyond tourism. But the critical method also sharpens our view of our domestic contradictions. What does the alpine pasture innkeeper get from the summer tourism? When the last e-bike rider has downed his spritzer apple juice, is it worth looking into the cash register? Who benefits from the landscape maintenance carried out by the pasture farmers, and what could be a fair price for it? Alpine summers, farm vacations, autumn stays—these popular tourism offers have become the cash cows of alpine summer tourism, but at what cost?

Questions, questions. Not all are easy to answer. Moreover, it is not the intention of this book to inundate the readers with figures, but rather to stimulate to further thinking, because thoughtfulness can be a first step towards sustainability. Tourism has become an indispensable part of the lifestyle for many of us and a tool for individual self-realization. But after the holiday trip, what remains? Have you rested well and had a great time, learned something new, perhaps even about yourself—or have you simply ensured yourself a suntan while your bank account dwindles?

Tourism has become industrially organized travel, but it has not lost its fascination because of this—quite the contrary. Never has there been more travel than in our affluent society, where more and more people can afford the temporary change of location, and set off in pursuit of short-term happiness in far-off lands. We travel following our dreams and longings—only to find outselves sniffing the air of paradise together with hordes of others.

The summit of pleasure beckons: travelling along the Via Culinaria, experiencing gustatory and olfactory highlights, and indulging our senses in that excess of wonder that a sunset with an alpine panorama or Zorbas music in the background creates part of a gourmet menu on the hotel terrace. Only on the backstage of the tourist hustle and bustle do the worrylines of an industry that is in fierce competition for customers and employees show: the need to react flexibly to shifting fashions, moods, economic cycles, and social conditions, to technical challenges and cultural changes in order to be successful. And there is the weather: too little snow, then again too much rain—and the effects of climate change are increasingly noticeable. All these factors affect the cost positions, but cannot be managed by business optimizations, IT solutions, or sophisticated management and marketing strategies.

Tourism is not just a sunshine topic, as suggested by posts on Instagram and Facebook, press releases, or countless contributions in glossy magazines. If the value created does not grow, simply attracting more guests is of no use. The industry needs permanent solutions and guarantees of continuity. The various influencing factors must be seen in a larger regional and global context. In the long term, the industry must commit to the Agenda 2030 of the worldwide climate agreement, which aims at network ecological, social, cultural, and economic goals, to be realized within the framework of *sustainability*. Sustainability is more than a question of safeguarding natural resources: the aim is to achieve lasting stability of ecological, economic, and socio-cultural conditions.

As in any other industry, there are also downsides. The intention of this book is to explore such contradictions, to highlight and analyse some of them, and to put them into a larger context. With an understanding of the interrelationships, a window of opportunity is opened to a more aware tourism—still experience- and pleasure-rich, but a tourism that is sustainable and can be operated without apology to future generations, even in times of climate change.

The advertising campaign launched by the Austrian National Tourist Office for the winter of 2021/22 featured the slogan *Winter Love*, with the Salzburg region on a first-name basis with the season and titles—almost in the sense of an invocation, *Dear Winter*—to attract guests for a stay in the ski areas. Couldn't *Dear Climate* be the forward-looking address for a multi-year all-season campaign?

Personal postscript

As the head of the Division of Transcultural Communication at the Department of Communication Science at the University of Salzburg, I have been dealing with the broad interdisciplinary and intercultural subject area, which used to be called 'foreigner traffic' and then became the more elegant and cosmopolitan 'Tourism'. In this field, I have been active in both practical and conceptual terms, as the leader of development projects with tourism components and as a member of the supervisory board of the Salzburg State Tourism Company. I have become familiar with the nitty-gritty details and the major challenges in tourism. Analytically, I seek to explore the cultural and communication-theoretical conception of tourism with empirical research. Given my deep commitment to development policy, the ecological dimension and the principle of sustainability have increasingly moved into the foreground.

In my presentation of the explanatory contexts, I therefore deal with those topics and regions that I know from my own work and in which I have considerable expertise.

Over the years, I have published a range of project and research reports, scientific articles, and books. Many of these earlier assessments have been confirmed and now serve as valid empirical bodies of knowledge. I myself often refer back to many of these texts to develop viable new theses. Scientists do write from time to time and, in the words of the great German humourist Heinz Erhardt, mostly by copying others' work. Therefore, I see no problem in admitting that I have copied from my earlier publications and cite them as references. This also applies to all other sources on which my insights or findings are based. For the most part, these are publications by friends and colleagues to whom I would like to express my gratitude.

I dedicate this book to my wife Karin, the wonderful companion on many journeys near and far, and through the thick and thin of life.

A final word on the English-language edition
The manuscript for the German edition of this book was submitted to the publisher in the summer of 2021, at the height of the Covid-19 pandemic. The tourism industry, which was hit harder than other sectors by the global restrictions on mobility, was expected to rethink its position. Today we must ask: Has this time been used to rethink and adapt to climate change? Will tourism be able to play its role more sustainably in the near future? How long will it take for it to be in line with European climate targets?

Probably a very long time, because while global air passenger numbers have reached a new high after the pandemic—4.3 billion in 2023—investment in sustainable aviation fuel (SAF) does not seem to be keeping pace. These and many other emissions are unlikely to decline any time soon, while visitor numbers to international tourism hotspots are already above the levels that led to overtourism before the pandemic. Increasingly, local residents are fighting back—protesting against the appropriation of homes for tourist rentals via Airbnb, for example, and calling for massive restrictions and controls, as well as better regulation of tourist flows.

Given the large numbers of tourists, we will all have to accept that individual freedom to travel will have to be curtailed to some extent. Even now, it is hardly possible to visit a museum or an interesting exhibition on the spur of the moment: early registration and booking are required. The same applies to busy old town centres and World Heritage Sites, in order to guarantee the quality of the visit as well as the quality of life of those who live there. Wherever the mass of tourists becomes a problem, stricter rules can be expected. Tourists will need to plan their trips more carefully. Tourist destinations will have to adapt their urban and transport planning. Popular destinations will have to reduce private car traffic and expand public transport—a sensible way to adapt to climate change. And tourism will need to be better integrated into urban planning.

Several cable car companies in the valleys of the Salzburger Land are already doing this in an exemplary manner. In the Alps, cable car companies have been at the forefront of the historic transformation from farming villages to tourist destinations, and they continue to be a dominant economic driver. However, they have recognised that their economic success must go hand in hand with the long-term protection of nature. Some companies have embarked on sustainability processes in line with the European Union's Green Deal. To reduce their environmental footprint significantly, they are increasingly switching to non-fossil fuels, generating their own energy from renewable sources and working to expand public mobility schemes.

This is the beginning of a rethink and a new transformation process that must involve the entire tourism industry in order to meet climate targets and make tourism a truly *green* business.

Kurt Luger

Contents

1 Always Home or into Happiness—Why Do We Travel? 1

2 Near and Far, in Between Longing 21

3 Places of Happiness, Mobile Privatization and Emotional Geography 49

4 A Brief History of Travel and Tourism 69

5 Utopias and Dystopias—Dream and Nightmare 91

6 The Past Has Never Been as Beautiful as Today 113

7 Alpine Tourism: Fair Weather Zone in Climate Change 147

8	Tourism as a Development Perspective	179
9	Paths to Sustainability	205
10	The Vision: Smart Tourists, Minimally Invasive	235

1

Always Home or into Happiness—Why Do We Travel?

Opinions about the purpose of travel will not always coincide,
as great witnesses have noted.
One should travel simply in order to travel, Goethe thinks, not
to arrive,
Enthusiasm, strictly speaking,
is the most intrinsic gain.
Montaigne saw returning as the sole purpose of travel.
Also Novalis emphasized this: Wherever we go, we are going
home.
But Seume, who walked all the way from the German Lands
to Syracuse,
Held that things will go better mentally if we walked more.
Eugen Roth, *Der Sinn des Reisens* (The Meaning of Travel),
a rhyming poem in the German original

All the misfortunes of this world come from the fact that people cannot stand being in their homes in the long run—so Blaise Pascal summed up the whole problem of our restless times. While Novalis still thought that every journey ultimately leads home, because life itself is to be understood as such, the modern lifestyle finds its goal in being on the move. *Out, just get out*, that's the motto. Mobility becomes a value in itself. But is it possible, to find ourselves while on the move? How can we expect to find ourselves when we are surrounded by attractions and thousands of distractions, tailored as pre-packaged products, so that all we need to do is accept them—and are then immediately catapulted into a state of happiness?—or so the hype goes.

The UN World Tourism Organization recorded around 1.5 billion tourist arrivals worldwide for the year 2019—as yet, the last year of the booming tourism economy. Estimated revenues from tourism for the year 2018 were calculated at USD 1.450 billion. In 2020, due to the COVID-19 pandemic, there were one billion fewer international tourist arrivals: this collapse led to an almost unimaginable loss of income of $1.3 trillion. This affected not only the big actors, but also the 100–120 million employees in this industry.

Tourism has become a major economic sector worldwide. To take one example, some 6% of the gross domestic product in Austria is attributable to the tourism and leisure industry. In the mountain regions of the Tyrol or Salzburg, tourism has become one of the most important economic sectors.

Until the COVID-19 pandemic put a temporary end to this development, revenues were increasing steadily, despite fears of attacks and terrorist attacks. Only a few years after the deadly attacks on tourists in Luxor and Bali, after the tsunami in Thailand and Indonesia, the fears were already forgotten, and revenues reached new heights.[1] Even plane crashes, mass traffic accidents on the way to holiday paradises or mega traffic jams have now become accepted as unavoidable side-effects. Many travellers have become accustomed to the increased danger, although, compared to previous epochs, travel has become safer. Ever since the medieval quest, the heroic journey of chivalry, the ultimate purpose of travel has been the happy or healthy return after passing the tests of endurance. Nothing has changed in this respect to this day.

What drives people out today, out of their comparatively privileged living conditions? After all, it is primarily the denizens of the wealthy Western industrial societies who set out. And they do this for *pleasure*—not because they, like migratory birds or nomads, can no longer find food, or have had to flee their homeland and seek asylum. Bumper to bumper, Germans and Scandinavians stream from north to south, from Stuttgart to Munich and further through Austria via the Tauern motorway to the coveted South. Brits endure hours of waiting at airports due to airspace overloads, in order to roast at some Mediterranean Aloha Beach; camping holidaymakers endure ceaseless rain on relationship-promoting six square meters of caravan space or in a survival dome tent. Surely, they would not accept the many inconveniences of leisure travel if there were not some overriding benefit that reduces these hardships to anecdotes and episodes, to conversation material for a few hours of small talk.

The motivation to travel, or the motives for setting off, form the key to understanding this global need for movement. Here we should distinguish between *push factors*—the internal driving forces—and *pull factors*, the external influences. Whether it's about compensating for various shortcomings, the recovery and maintenance of one's own body, the search for beauty, expanding our intellectual horizons, or about finding expression for suppressed need—all travel decisions are influenced by motives of some kind.

The *push* factors tell us that we have to get away to meet shortcomings at home—sun instead of drizzle, relaxation instead of hustle and bustle, etc. *Pull* factors determine where we travel—why we are drawn to a certain place. Basically, we want to get away, no matter where the next plane takes us—it could be anywhere. And the tourism industry advertises tirelessly, invents trends and fashions, lures with prices or the uniqueness of holiday destinations—making them a 'must'. And then, the next year another destination beckons, and the caravan moves on.

But there are also holidaymakers who have visited the same place for decades, and want to go only there and nowhere else—perhaps because of the enchanting landscape or the size of the schnitzel portions, or maybe because of the landlady in the dirndl dress or because sleep in the hotel's own pinewood bed provides complete relief. This state of satisfaction, of affordable happiness, is repeated annually. Such regular guests take no risks, because they know what to expect. Indeed, after 25 years, their loyalty to the Pension *Waldbrand* (Guesthouse *Burning Forest*) is rewarded with an honorary pin from the municipality: now the guests can feel a bit like locals.

In the past, only a few traveled—for professional reasons and to earn a living, the craftsmen; for reasons of courtly education and intellectual training, the young nobles on their Grand Tour, who visited other royal houses, the main churches and libraries—probably also the most notorious brothels or dives. The gradual social and political transformation of society brought about a democratization of travel. In the middle of the nineteenth century, pleasure travel for the well-off emerged: the first travel agencies were established, and the bourgeoisie moved from the cities to rural summer retreats. In the Western world, the generally peaceful period after the Second World War and the increase in wealth and leisure time led to the integration of holidays into the lifestyle of industrial society. Holiday trips now became possible for practically all parts of the population.

This temporary exodus of the masses from their busy, work-oriented lives also led to criticism of this society, which built up such massive deficits that they could hardly be compensated in three weeks of recovery.[2] But the world of work was changing, and with the emerging post-materialism also the hierarchy of values, paving the way for the 'leisure and experience' society.[3] Today people work to live as well as possible, the 'away-from' needs gradually giving way to 'towards-to' needs. Self-realization, freedom, maximization of pleasure have a different place in the hierarchy of values in today's Me-Society for leisure and for work. The self-determination experienced in leisure time or on holiday is also demanded at the workplace, and tends to replace external determination. People long for immediate sensual experience; they seek social contact and authenticity, want to feel the ground under their feet and not just function at their jobs in the industrial plant, in the office or in the kitchen. Many are seeking self-improvement, completeness, and to strengthen their self-esteem.

The Fun and Experience Society

Sociologically speaking, all this is a kind of contrast experience. The spatio-temporal distance to everyday life and the demonstrative consumption of experience away from home convey the impression of being in the middle of pulsating life, fully involved and participating in the achievements of modernity or the affluent society: this is the case with the holiday trip for the vast majority of people in Western industrial societies.[4] The contrast to everyday life is the attraction: it breaks with our accustomed rhythms, and compensates for the gruelling monotony of everydayness. Now, in times of Covid-19 and its more dangerous variants, people long for normality: they want their old life back, the everyday life with its ingrained rituals and conventions, but also the minor escapes and pleasures, which are suddenly not possible due to curfews, restrictions on mobility and distance regulations. Not being allowed to travel is widely perceived as the most massive restriction of all.

With a successful holiday at a prestigious location, we can even boost our status, at least in the struggle for recognition among neighbours and relatives. With each postcard—that illustrated medium of a beautified world, which began its triumph in tourist communication as a correspondence card in 1869—this increase is conveyed in writing. The heartfelt greetings—now mainly sent via Instagram, WhatsApp or Facebook—announce that our everyday frustrations have been set aside at a mountain lake or on a Mediterranean beach, while our damaged identities, if not our souls, have been successfully repaired.

Bardolino and Moonlight, in the Sinful Night

Tourists not only want to recharge: they also want to experience something, bring variety into their lives. This need—a change of scenery—also indicates a state of psychological saturation. It is not physical fatigue itself that triggers a drop in performance, but the discontent and reluctance that accompany it. A holiday trip is expected to bring about the necessary change, boosting the level of satisfaction. Moreover, where no one knows you, you can live a little more unrestrained: party-goers of all countries, unite! The control of emotions is lost, or at least much weaker, in the leisurely state of exception on the Balearic or Canary Islands; travel advertising conjures up the contrast to everyday existence and spurs new needs and expectations: alcohol excesses, the indulgence of sexual preferences, unrestrained hedonism, loss of the sense of space and time and in many cases probably also of decency—all inclusive! In this, today's tourist travel is somewhat similar to the role formerly played by annual festivals, funfairs and rituals, a brief escape from the usual order as a form of emotional venting.[5]

By detaching themselves from their familiar environment, becoming someone else, tourists free themselves like masked carnival participants from constraints and identity pressure. The rules are not set aside completely—but in swimming trunks on the beach, social differences are levelled; and in the mountain huts, everyone is on first-name terms, as if the classless and prejudice-free society had long since become a lived reality.

In the limitation of one's own self and the desire to escape the familiar environment, there is probably an answer to the question of why this periodic departure from everyday life and the return to it has become a cultural convention and a global pattern, from which a global market called 'tourism' has developed. We are given the opportunity to model dream and reality, to let imagination and real existence merge. Tourism thrives on enchantment, illusion and longing, from staging—and also from manipulative seduction dishonesty. 'Experience marketing' aims to convert entertainment desires and curiosity, needs for boundary crossing, renewal—yes, even for redemption—into consumable products and economic performance.

Tourism professionals and their marketing experts are veritable artists in storytelling, in inventing traditions and meanings: they codify destinations as 'worth seeing', even culturally overwriting entire landscapes: At this lake

Beethoven or Schubert mentally composed symphonies, on that quiet meadow the young Hans Krankl (Austrian football superstar of the 1970s and 1980s) practised his kicks; in this backyard David Alaba (current Austrian football superstar) once dribbled; and everywhere where his majesty the Emperor once stopped to relieve himself, on the Via *Imperialis* (as the Austrian cultural tourism advertising campaign of at the time was called): even otherwise insignificant places could become attractions worth a visit, or at least a minor detour.

Through Europe, the Council of Europe has certified cultural routes such as the Mozart Way or the Via Habsburg. In journeying through several countries, visiting castles, fortresses, magnificent churches, monasteries, abbeys, universities and museums, along the significant places of monarchical conservatism, late baroque enlightenment and scholarship, can retrace a thousand years of history.[6]

There are certainly no limits to the thematic imagination of tourism professionals when it comes to storytelling. We can note the countless artificial leisure worlds with their individual entertainment guarantee, or the theme paths designed as tourist experience worlds in the context of regional culinary specialties, such as the Baden Asparagus Road, the German Hop Road, the Waldviertel Textile Road, the Upper Austrian Cider Road—and the countless wine roads.

Dreams animate enliven and colour our inner maps. This also applies to film and television. Like the real change of location on trips, they give space to the non-ordinary, create room for our curiosity and let our wishful dreams become reality for a few hours, days or weeks. We travel to boost our own quality of life, to get closer to our dreams of an imagined world, one more beautiful than at home.

Travel as a Key Practice of Lifestyle

In the late modern culture of performative self-realization, travel becomes a form of life design. We strive for singularity, for the special, cultivating the elements of life guidance, to make something special out of it, ideally a work of art. Andreas Reckwitz[7] has convincingly shown that the new middle class combines two formerly opposing ideals: the *romantic ideal of self-realization*, which emerged around 1800 and which includes sub- and countercultural designs such as the bohème or counter-culture, and the *bourgeois ideal,* which strives for social status through performance and education. It has dominated the middle class since the nineteenth century and still exists today. We want both self-development and social success.

This new middle class has high levels of education and culture, and lives in the mode of singularization, always in search of the special, the extraordinary, the 'authentic'. All cultural goods become singularity goods, co-creators of a lifestyle of self-realization and self-transcendence. Self-realization applies to the area of leisure travel as well as to the area of work, personal relationships, partnerships and family. The cosmopolitan-oriented late-modern subject lives in a hyperculture, in which the classic boundaries of the culturally valuable have been dissolved—the boundaries between the contemporary-modern and the historical, between high and popular culture—and also between one's own and foreign cultural circles. Whatever may contribute to singularity and self-development is considered, acquired and appropriated from the global fund of cultural goods.

In such a framework, the journey must contribute to the extraordinariness of the traveller, must lead to special places, far away from the catch-all holiday resorts of industrialized tourism: authenticity must be tangible. The sovereign self has competence, education and cultural connoisseurship to appreciate the quality of such a destination. Vacation and travel become 'intentional creative status work'; documented in pictures, they transform everyday life or its uniqueness into scenes from which something like 'perceived distinctiveness' arises through the social media. Influenced by the romantic tradition, where great emotionality, originality, diversity, poetry, picturesque places, unique peoples and cultural circles were central, today's late-modern subjects find self-realization with empathetic individuality. This lived emotionality in its pleasurable form moved into the centre of late-modern forms of life as a *leitmotif* that extends far beyond the new middle class. We strive for positive emotions, because joy, fulfilment, pleasure, enthusiasm, social harmony, intensity, unfolding of the self in all facets—in a word, *happiness as a sense of life*—have become as a central driving force.

However, there is a built-in risk of failure, in the form of 'negative unavailabilities'[8]: disappointments in the face of the perceived discrepancy between expectation and reality. Frustration, fear, anger, sorrow are emotional reactions to defeats—and they lead to various dilemmas and call for coping strategies, because an exit from the culture of self-optimization, of societal progress and the pursuit of happiness as an individual project is in fact not intended.

Life Passes as Quickly as a Vacation

On the bus to the train station: I offer my seat to an elderly lady, we start a conversation. She tells me that she is almost 90 now, can't remember this and that anymore, and she has the impression, that life passes as quickly as a vacation ….

Compared to earlier epochs of societal development, today's life in a society striving for prosperity and individual benefit is subject to increasingly rigid temporal rules. This high-speed society has said goodbye to many of the promises of modernity such as greater autonomy or self-realization. According to the German sociologist and political scientist Hartmut Rosa, it has caught itself in an acceleration circle that drives our pace of life.[9] However, this social acceleration allows less and less realization of life plans: the political shaping of society is less and less in line with ideas of justice, progress, sustainability. Rather, ever-accelerating societal development leads to a *disembedding of space and time*, as diagnosed by Anthony Giddens.[10] It produces a growing alienation of the individual from his spatial and material environment, because it hardly succeeds in combining action and experience episodes into a whole life. Jürgen Habermas[11] called this the *triple alienation of the ego*—with society, with inner nature, and with outer nature. The ephemeral world of the twenty-first century allows fewer and fewer resonante relationships, and increasing self-alienation and world-alienation are the consequences.[12]

The Vacation as an Antithesis to Acceleration

Contrary to this time dictate, holidays or tourism seem to stand as part of the promise of modernity, despite remaining trapped in the social norms of temporality that generate time scarcity in late-modern societies. It is part of the system in the sense of a repair operation, so it cannot escape the time-economic calculation. But it also falls out of time, because the amount of time available (guaranteed also in the sense of the legally anchored holiday entitlement for employees), is greater, and it is up to the individual to decide how to spend this time.[13] Like the *adventure*, which holds its extraordinary status only when contrasted to bourgeois everyday life, so working time and holiday time remain inseparably connected.

The time dictate also applies to the limited time available for holidays: the days seem to fly by twice or three times as fast as in everyday life, because we want to live particularly intensively, or in fact bring life into our lives. The intensity of the experience makes time shrink faster. Sociologist Niklas Luhmann[14] held that time is not scarce in itself: the impression of scarcity arises from the overload of experience by expectations. What applies to everyday life continues on vacation. The originally social construct of time has become autonomous, confronting us as an independent force that we handle as a resource under purposive-rational and time-economic aspects. Any time-gain is immediately processed and used rationally again, and even more experience is crammed into a time reserve. Thus, time scarcity arises as a result of a rationally perceived handling of time—not as a result of incompetent time management.

All too soon, the holiday-maker is on the plane back home, savouring the afterglow of the experience. Our high-speed society demands a temporal condensation in the appropriation of the new and strange, a compressed consumption of experiences. A study among German-speaking visitors to the Third World showed that respondents found it very odd that the people living in these countries seemed to have time in abundance.[15] The men would sit in the coffee houses during the day and play cards, while the women chatted by their house doors. Speed was not evident. That may be a superficial impression, as the actual work processes of the local people usually escape the tourist eye. But it is clear that cultures live at different speeds. According to Andreas Obrecht, there is time scarcity in the centres of economic activity because the future is realized in the pace of the present, growth and productivity increases are tied to the systemic scarcity of time. Acceleration and time economy have been guarantors of added value and wealth since the industrial revolution. In poor societies, on the other hand, there seems to be time in abundance, because the present is not a transitional stage for the goals to be achieved in the future. Therefore, in the centres there is wealth and time poverty—on the periphery, there is poverty and time wealth.[16]

Slowing down, dreaming of islands, retreating into deceleration, according to the motto, *Better a little movement in Jamaica or in Upper Pinzgau, than hurry, hurry in Germany*—this is the stuff of the longings that are cultivated by millions of relaxation addicts over the year who then let off steam for two or three weeks. The travelling self finds refuge in foreign places and thus experiences a temporary exit from a world of impositions, demands, and overstrains. This escape from everyday life, from time as well as from space, cannot comprehensively explain vacation and travel behaviour. But escapism[17] is central. Along with the curiosity for new things and breathtaking beauty, it is among the central driving factors behind the recent growth of the leisure industry.

Many people today are experiencing the idleness of leisure in their final years—or on holiday. The ambulatory repair operation of tourism has also developed offers for this, catching the driven ones of the chronocratic world with offers that promise moments of 'purposeless' experience, that offer slowness or speed reduction as compensation or therapy against restlessness and the rat race.

The global tourism economy—as fast and flexible as it shows itself in its marketing and the hunt for customers—corresponds to the slow chronotype in terms of pursuing the UN development goals (SDGs, Sustainable Development Goals) or its own sustainability goals. It has taken decades for the principle of sustainability, starting from Rio 1992 and the resolutions of the UN General

Assemblies, to enter the calculations of the industry, into tourism policy or into tourist practice at the national level or to find implementation. In several sectors—foremost among them cruise shipping and low-cost air travel—we will probably have to wait even longer, because their short-term or particular economic interests are contrary to the objectives of the global Agenda 2030.

Time in Happiness

Tourist habitats are *heterotopias*, or *other-places*.[18] We experience them as spaces of desire and as places of happiness.[19] The management of experiential spaces by tourism turns cities and industrial areas, but especially natural spaces, into places of tourist significance, which are sensually explored and emotionally appropriated. In the world of imagination, these are earthly paradises, filled with luxury, with culinary highlights, with beautiful women or handsome men; in winter they are five-star wellness landscapes and avalanche-free powder-snow slopes. Atmospheric sunrises and sunsets as well as glimmering peaks are also part of this. These images are circulated millions of times in the media and advertising brochures, luring us with images of a tempting emotional geography.

Beyond the stagings of media happiness, which includes a 'successful' vacation, there is also the path inward, to oneself, as a quiet form of approaching bliss. Tourism lives from this experience of an 'other-place' and 'other-time', which provides at least short-term disengagement from the strictures of everyday time. But it is not an experience you can take home as a souvenir. What remains is the enchantment of the place, the excitement of a different way of life, a different pace and the amazement at how long a dinner (or the wait for it!) can take.

Travelling at Human Pace—on Foot to the Self

The return to oneself and at a human pace, in hiking or walking, has experienced a massive surge and also a change in meaning in the form of trekking or hiking tourism. Finding one's own pace, searching for meaning on new paths or old trails—hiking tourism has not only long traditions in the mountain regions of the world, but is also totally in line with the trend in connection with personal responsibility for health.[20] The act of walking as a self-experience of a sedentary culture becomes a technique for dealing with oneself.

Mountaineering is an attempt to overcome the constraints of Western society through physical movement in the vertical, while allowing the body a challenging space for development. Similarly, the intended reconciliation with nature and the search for a simple life explain why the cultural practice of pilgrimage, prescribed by faith among Hindus and Muslims today, is increasingly found in our secular societies. The *homo viator*, the modern, ever-searching human, has grown out of this niche market. The peregrination has evolved from a penitential walk with a religious background—also Christianity has a long tradition of pilgrimages—to a foot journey in an open, spiritual tourism. It is fed by the longing for post-materialistic values, holistic ideas and a craving for the super-natural, with spirituality seen as a life attitude seeking meaning and significance that goes beyond the self. The slow, physically probing movement of forest bathing, for example, allows immersion in a counter-world of natural conditions. The landscape of trees and plants becomes the object of sensual experiences and moments, in contrast to faster modes of transportation and the *zeitgeist* of racing.

Time wealth allows us to slow down the pace in order to find intensity. *Slow Tourism* points the way towards nature and culture proximity or sustainability, because the quality or intensity of perception and experience depends on the way and speed at which people move through space. If the big city leads to an 'intensification of nerve life' and a 'condensation of time', as the German cultural philosopher Georg Simmel[21] formulated it, then the experience of nature and tranquillity is the complement for a fulfilled life. The tradition of summer retreat springs from this city/country contrast and with it the rise of the region around Semmering (Viennese Alps) or the scenic Salzkammergut lake district to destinations of longing.[22] Anyone who longs for something is in fact experiencing a lack of it, because longing denotes a state of undersupply, which expresses itself emotionally and physically as feelings of deficiency and which demands satisfaction.

Many Alpine tourists come from urban agglomerations, where nature is often reduced to roadside greenery and plants that resist greenhouse gases and heavy doses of dog faeces. It is also empirically proven that people living in rural regions travel less compared to city dwellers. The founding of nature and alpine clubs in the mid-nineteenth century was carried out by nobles and the educated bourgeoisie of European metropolises.

Nature-based tourism in protected areas or national parks aims to make the qualities and values of an area visible and to enable visitors to understand them and have pleasant experiences of their own. The concept of nature enjoyment dates back to the father of physical geography, Alexander von Humboldt (1769–1859) and encompasses a spectrum of experiences that are not strictly necessary for life, but that enrich, beautify, fill our lives with meaning and increase our *joie de vivre*. This can bring us closer to nature—as the German philosopher Ernst Bloch called it—as a wholeness of life.

Frozen Time—Photography as a Metaphor for Standstill

An adventure cannot do without a narrative about it; the same applies to holidays or tourist travel, which must be documented in order to capture for posterity the highlights of the year. In the narrative approach and the unfolding discourse with oneself, the chance for self-affirmation arises, whereby experiences can be anchored in one's own biography. Analogously, it is photography or the photographic inventory of the foreign that makes the holiday a memorable event.

Whereas professional photographers try to 'catch the decisive moment from life in a snapshot', amateurs are mainly interested in taking private memory photos.[23] The whole world in a selfie[24]—it's a matter of capturing the fleeting and transient, putting the perishable on ice and thus preserving time, preserving a past state as a desired memory in photo albums or digital storage.

Photographing in the case of special circumstances such as being on holiday fixes not only a slice of reality, but also a slice of time, countering the disappearance of time. Holiday photographs provide views of times, situations and events long past, bringing people and objects from their spatial and temporal distance into the present, and preserving disappearing events for posterity.

Every photograph—as Susan Sontag writes—is therefore a kind of *memento mori*. "To photograph is to participate in the mortality, vulnerability, and mutability of other people or things. Precisely by slicing out this moment and freezing it, all photographs testify to time's relentless melt."[25]

Holiday snapshots are memories of a temporal state of exception, of beautiful hours, days or weeks, an attempt to recall events or to force time to stand still—an attempt to hold onto a handful of happiness for eternity, because the holiday is over so quickly, almost as quickly as is life itself ...

Endnotes

1. Wolfgang Aschauer, *Tourismus im Schatten des Terrors*. (Tourism in the Shadow of Terror. A comparative analysis of the effects of terrorist attacks—Bali, Sinai, Spain), Munich: Facultas 2008.
2. This analysis is most clearly expressed in the publication of the Swiss economist Jost Krippendorf, *Die Ferienmenschen* (The Holiday People), Zurich: dtv 1984.
3. The change in values and with it the change in lifestyles is discussed in Roland Inglehart, *Modernization and Postmodernization*, Princeton University Press, 1997; Gerhard Schulze, *Die Erlebnisgesellschaft* (The Experience Society), Cultural Sociology of the Present, Frankfurt: Campus 1992.
4. Horst Opaschowski, *Umwelt.Freizeit.Mobilität* (Environment.Leisure.Mobility), Wiesbaden: Leske + Budrich 1999.
5. A very apt and still valid overview of tourist behaviour is provided by Christoph Hennig in his *Reiselust* (Wanderlust, Tourists, Tourism and Vacation Culture), Frankfurt: Insel 1997.
6. https://www.coe.int/de/web/cultural-routes/via-habsburg, https://www.coe.int/de/web/cultural-routes/certification, accessed 21 August 2021.
7. Andreas Reckwitz, *Die Gesellschaft der Singularitäten* (The Society of Singularities). Berlin 2017; *Das Ende der Illusionen* (The End of Illusions, *Politics, Economy, and Culture in Late Modernity*), Berlin: Suhrkamp 2019.
8. Hartmut Rosa, *Unverfügbarkeit* (Unavailability), Vienna: Residenz 2019.

9. Hartmut Rosa, *Beschleunigung und Entfremdung* (Acceleration and Alienation—A new theory of modernity), Berlin: Suhrkamp 2013.
10. Anthony Giddens, *The Consequences of Modernity*, Stanford University Press, 1990.
11. Jürgen Habermas, *Zur Rekonstruktion des Historischen Materialismus* (On the Reconstruction of Historical Materialism), Frankfurt: Suhrkamp 1974.
12. Hartmut Rosa, *Resonanz (Resonance—A Sociology of our Relationship to the World)*, Berlin: SV 2016.
13. For the distinction of time budgets according to determination time, obligation time and disposition time, see Horst Opaschowski, *Freizeitökonomie—Marketing von Erlebniswelten* (Leisure Economy—Marketing of Experience Worlds), Opladen: Leske+Budrich 1993.
14. Niklas Luhmann, Die Knappheit der Zeit und die Vordringlichkeit des Befristeten (The scarcity of time and the urgency of the temporary), in: *Die Verwaltung—Zeitschrift für Verwaltungswissenschaft (Administration—Journal for Administrative Science)* 1/1968, 3–30.
15. Bernd Schmidt, *Der Orient—Fantasia 1001 Nacht* (The Orient—Fantasia 1001 Nights. How Tourists See and Understand the Foreign), Ammerland: Studienkreis für Tourismus und Entwicklung 2001.
16. Andreas Obrecht, *Zeitreichtum-Zeitarmut. Von der Ordnung der Sterblichkeit zum Mythos der Machbarkeit* (Time wealth—time poverty. From the order of mortality to the myth of feasibility), Frankfurt: Brandes&Apsel 2003.
17. This term is used in communication and media studies as an explanation for the retreat from the overwhelming pressures of everyday life and the escape from the real world into the world of images of entertainment and distraction media. In tourism studies, it is considered a travel motive or a justification for the temporary departure from the routines of normality. Concerning tourism, see for example Jörn Mundt, *Einführung in den Tourismus (Introduction to Tourism)*, Munich: Oldenburg 1998.

18. For an overview of spatial theory, see Stephan Günzel, *Raum: Eine kulturwissenschaftliche Einführung* (Space, A cultural science introduction), Bielefeld: transcript: 2017; and Marc Augé, *Nicht-Orte* (Non-Places), Munich: Beck 2012.
19. For more details, see Karlheinz Wöhler, *Touristifizierung von Räumen* (Touristification of Spaces), Wiesbaden: VS 2011.
20. Gabriele Knoll, *Handbuch Wandertourismus* (Handbook of Hiking Tourism), Konstanz: UTB 2016; Christian Hlade, *Das große Buch vom Wandern* (The Big Book of Hiking), Vienna: Braumüller 2019.
21. Georg Simmel, Die Großstädte und das Geistesleben (The Metropolis and Mental Life), in: Georg Simmel: *Gesamtausgabe: Aufsätze und Abhandlungen 1901–1908)*, Frankfurt: Suhrkamp 1995, 116–131.
22. Wolfgang Kos (ed.), *Die Eroberung der Landschaft* (The Conquest of the Landscape. Semmering-Rax-Schneeberg). Catalogue for the Lower Austrian State Exhibition Gloggnitz Castle 1992, Vienna: Falter 1992.
23. Thomas Theye, *Der geraubte Schatten* (The Stolen Shadow. Photography as an Ethnographic Document), Munich: Bucher 1989.
24. Marco d'Eramo: *Die Welt im Selfie. Eine Besichtigung des touristischen Zeitalters* (The World in a Selfie. An Inspection of the Tourist Age), Berlin: Suhrkamp: 2018.
25. Susan Sontag, *Fotografie* (On Photography), Frankfurt: Fischer 1999, 21. (original: New York: Farrar, Straus and Giroux 1977).

2

Near and Far, in Between Longing

Do you know the highest mountains?
I believe that they are the longings of people ...
H.C. Artmann

I haven't been everywhere, but it's on my list.
Susan Sontag

The tension between proximity and distance, the familiar and the supposedly foreign, the known and the imagined: this dialectic is as inherent to tourism as the bow wave is to a ship. When we travel, all our antennae are out, always looking for differences from the familiar. The approach to the foreign begins with constant comparison: "This bay looks so much like the one in Portugal!", "They make pasta in a completely different way than we do!", "You wouldn't see such filth at home!", "These leather jackets are much cheaper than ours and so chic, let's take two!". Our own frame of reference, our own cultural order serves as a frame of reference, the familiar as a measure of things. It has to be this way at first, because travellers need a point of view, especially when they are moving so fast through space that they see nothing or only the outlines of an impenetrable reality.

Exoticism and Xenophobia—the Allure of the Foreign and Its Rejection

The contrast and thus the inevitable attraction of the familiar and the foreign is deeply rooted in human nature. Ethnopsychoanalysis traces it back to human stages of development. Fear of the foreign but also fascination—both are in us and are taught to us. A xenophobic basic mood encounters not only other cultures with hostility and rejectiony, but also those areas of our own culture that are "different": Freud referred this phenomenon as the "inner foreign country", which can threaten identity.[1] Julia Kristeva holds that foreignness resides within us: we carry the basis of our behavior towards the foreign in us, because "we are strangers to ourselves".[2] Fear fantasies, projected onto foreigners, acquired through socialization and personal experience or conveyed by the media, spur the production of enemy images. Latent in every traveller, they are activated and articulated, especially where one's own symbol and meaning system fails, indecipherable due to sheer foreignness.

While xenophobia is rooted in child development and socialization, exoticism is a phenomenon of adolescence. Youth, the time of forced identity formation, of explorations and experiments with breaking out from the protective circle of the family, is a time of world discovery and self-experience. The foreign is fascinating, attractive and desirable. New geographical and sensual horizons open up, the world expands, and self-experience brings the self to life. In exoticism—in the sense of youthful curiosity—there is a significant driving force for tourism in general.

The early voyages of discovery and to the first circumnavigations of the globe were undertaken by risk-seeking sailors, mercenaries and merchants under the flags of their kings and queens. They sailed because of expectations of the high profit from the trade in spices and silk, combined with greed for precious metals and gems. Accompanying scientists like Louis Antoine de Bougainville or Georg Forster tried to trace the blueprint of the world and its biodiversity. Their travel descriptions contain numerous references to conditions that we would interpret today as Elysian fields and exotic idyll. After months at sea, the Europeans were fascinated by the uninhibited lifestyles of the natives, the symphony of intoxicating smells and the botanical abundance of the tropics. It seemed to them as if they had sniffed the air of Paradise itself. Sexual freedoms entered the picture, encouraging other daredevils to overseas adventures and a bold life.[3]

When Central European tourists today venture further than the upper Adriatic, many probably also want to immerse themselves in this ethnically colorful pageant of exoticism. For more reflective natures, such journeys can lead into the self and into greater thoughtfulness. They can change ways of thinking, because—as that master teacher Goethe formulated it in his *Elective Affinities*—"no one walks unpunished under palm trees, and one's sentiments certainly change in a country where elephants and tigers are at home".

Interrail, youth trips or language holidays abroad are offers of the tourism industry, but they also serve as rehearsals in a society that has made the crossing of one's own borders and experience of foreignness—albeit industrially organized—into a constitutive and appreciated part of the cultural order. Escape from everyday life and the confines of the family, away from the familiar and towards the extraordinary, with experiential intensification and elevation of the moment—these are the dominant basic motives of any tourist undertaking.

Professional Exotic Management— Warmth in the Far Distance

Christoph Hennig writes in his book about travel lust and holiday culture that tourist travel is rarely about seeing something completely new. Rather, we hope to experience the truth of our collective fantasies. Tourism unfolds in the tension between culturally mediated fantasies and real change of location. Its goal is a seemingly paradoxical form of experience: the sensual experience of imaginary worlds.

How much 'foreignness' can tourists take? Intercultural communication theory holds that fear and uncertainty are the determining factors for the extent of what is bearable, leading back to the psychogenesis or sociogenesis of the relationship with foreignness. The less one knows about the foreign and unknown, the greater the fear will be, the more reserved the interaction, and the smaller the scope of action. Even today, only a few percent of Germans or Austrians venture beyond the familiar holiday destinations on the European coasts or in the Alps.

The history of summer tourism illustrates this. The first trips of German tourists, who rapidly became motorized after the Second World War, led to areas where they had already spent their holidays before the war: to the domestic coasts, to the Austrian and Bavarian mountains, to the lakes of the Alpine foothills. "Not at home and yet at home" *(Nicht daheim und doch zuhause)* was a slogan of the Austrian tourism advertising in the 1970s, which successfully flirted with the familiar in the foreign. Over the years, the radius of action expanded beyond the Adriatic, as the coasts of Spain, and the Greek islands, later Turkey and Tunisia, were added. This was due to transport-related reasons such as the development and reduction in cost of air travel, but also to the rapid increase in individual mobility. Thanks to professional tourism marketing, low prices and favourable exchange rates, and with popular songs and films that promoted Austria and Italy as ideal holiday destinations, the Alps and the touristically transformed Mediterranean area soon became familiar terrain. During the 1980s, more distant destinations increasingly came into focus. Dream destinations in the Third World became more and more popular, as dreams of great freedom and the mastery of the tourist situation in now-familiar foreign countries spurred the curiosity of travellers for new tourist challenges.

It is not only the travel industry that conducts professional management of the exotic. The media—from travel literature to television magazines, from homeland films to slide presentations, millions of websites and postings; the entire communication and cultural industry—are lively actors in the business of travel longings. Vacations and short and long-distance travel have become a coveted part of the bourgeois lifestyle.

Travel journalists, every blogger or influencer and travellers sending snapshops on their smartphones can thus become helpers of escape into real and false paradises. Tourist photos reflect the ancient human dreams of peace and abundance, of friendship and happiness. They construct myths of the power of nature and of the magic of an encounter with simple Greek wine and the accompanying peasant farmer. In countless articles or television images about the Maldives, Mauritius or the Canary

Islands, the same sunsets appear again and again, with pristine white beaches populated by almond-eyed native girls or dark-skinned young men who move like big cats. How often have the holiday clubs under the Southern Cross—although often walled-in islands of abundance in a landscape of poverty—been invoked as the epitome of the human dream of freedom? The agents of the experience industry are constructors of a beautified image of the foreign—which can also be found in every printed or digital travel catalogue and is offered there as part of the product for sale.[4]

Tourists rarely encounter a foreign culture, but rather its myths or stagings reduced to stereotypes. But as long as the performance does not seem off-the-peg, the staged authenticity is accepted: the participants already know that the "real" and unadulterated is a mystification, and the locals act as if they were on stage when they feel observed. Tourists and locals play their roles: both sides know the rules of the game. The tourist staging becomes a real fake—tourism lives from enchantment, seduction, illusion, it belongs to the entertainment industry like the Saturday evening TV shows, or the Hollywood films, in which the world is far more colourful than in reality. The attraction lies in the condensation of diversity, the choreography of highlights, which leaves one speechless, makes one marvel—although not necessarily understand, because there are whole worlds between the Rosenheim Bavarian carpenter or the Salzkammergut Gmunden ceramics dealer on a TUI Group vacation and the dancing Masai shepherd or the the subtly smiling beauty offering bananas in the hills of Nepal. There is enough difference to arouse surprise or enthusiasm, but this also has an experiential character, albeit fleeting or superficial. After all, tourists are not professional ethnologists. They usually content themselves with cultural trivialities marketed as 'worth seeing'. Artefacts such as the parade of the gold hood *(Goldhauben)* women of an Alpine village or the slap dance of the *Holzhackerbuam* (wood-chipper boys group) as the highlight of the Tyrolean dance & folklore home evening in the Salzburg mountains attract the attention of visitors with their cameras and smartphones.

By contrast, the *ethnologist* wants to penetrate the foreign culture, and understand it in its inner logic—and this requires learning to interpret the alien symbol system and its codes. To understand the culture of a people, according to the American anthropologist Clifford Geertz, one must reveal its normality without neglecting its particularity. Placed in the context of their own everyday life, their incomprehensibility fades, and becomes accessible.[5] For tourists, however, the culture of a foreign people remains essentially inaccessible, of interest mainly for specially interested travellers and hobby anthropologists.

Such an encounter cannot be organized as a package tour, because cultural encounters require time, a slow and gradual approach, until place, landscape, and what is seen lose their otherness, and people of different origin, ethnicity, religion, and cultural order can enter into a closer exchange relationship. Individual travellers may be able to penetrate more deeply, but only if they—as Peter Matthiessen suggests—approach the foreign with a calm expectation of the things that will come, free of defense mechanisms, with little luggage and without holding on and rejecting.[6] In his *Indian Diary*, the sociologist of religion Mircea Eliade admits that it is almost impossible to understand anything in a place where one falls out of the frame. For him, there is only one sure way to do justice to a landscape or an experience in Asia: namely *not* to look for anything specific. "If you're lucky, you'll encounter something completely unexpected—if not, try somewhere else. But there is no prospect of experiencing anything extraordinary and gaining a genuine understanding of the phenomena if you assume that you will find everything you are looking for. The reason is simple: Man learns nothing on his own, does not experience anything independently, everything is revealed to him."[7]

Escape Helpers into Paradises

The ethnological research report, the literary and the journalistic travel report have one goal in common: to transcend one's own cultural borders. But they achieve it or fail at it in different ways, because their tools are different. The task of ethnologists is the thick description, the interpretive penetration of a foreign culture. To do this, they must learn to think and feel like those they are studying: they must leave the position of outsiders and become involved as cultural nomads. Their background and cultural origin play a subordinate role, although these flow into their observations and into the text. As Ernst Bloch once noted, you take your own culture with you wherever you go. In the times of colonialism or the German Reich, ethnologists also took with them the racism prescribed to them nationally, had it confirmed on site, and disseminated it as 'scientific knowledge'.

The literary travel report, which has long traditions as a genre, gives the ego room to manoeuvre with much more leeway.[8] The basic pattern—"I was there"—they share with the ethnologists, but the ego becomes more important: filter that indicates a different direction. The readers should be able to understand what the alpine travelling authors have felt, should be able to share in their interpretations, without having to make a cultural change, a shift of perspective. Depending on intercultural empathy or perspective, the reader may learn more, or less, about the countries visited in these literary explorations—but very much about the writer.

Examples of the biographical form, in which the foreign culture is seen as a fascination as well as a discovery enriching one's own cosmos, include Goethe's *Italian Journey*, the philosophical meditations of Pier Paolo Pasolini on the *Breath of India* and Herbert Tichy's life-wisdom *What I learned from Asia*. Alexander von Humboldt (1769–1859), who relied heavily on empirical facts, figures and tables for the description and analysis of his observations in South America, stands paradigmatically for the observing form. Bruce Chatwin's or Cees Noteboom's sensitive travel narratives present encounters with people and their exotic ways of life. The comprehensive Tibet reports by Sven Hedin or Alexandra David-Neel are examples of highly demanding adventure literature. Like the explorations of the world-travelling merchant's daughter Ida Pfeiffer or the globetrotter Richard Burton, their journeys more than a hundred years ago led to then hardly known world regions. For the reflective tradition, the thought images of the *flaneur* Walter Benjamin are characteristic, because in them the observation of the cityscape is connected with reflections on the collective structure underlying it. The same applies to Simone de Beauvoir's travel diary, *America Day by Day*, which she published from her extensive America trip in 1947.

Heinrich Heine, the "last poet of German Romanticism", gave us the genre of the literary *travel picture*. His *Harz Journey* (1826) marks a paradigm shift in the field of travel literature by depicting the political stagnation of the Restoration period as contrasted with the ideological homebody of an entire epoch. The criticism of the conditions prevailing elsewhere was meant to refer to the situation in his own country, hemmed in by censorship. In his travel images, the restless search for his own place in history and his own restlessness are expressed—and this marks the beginning of literary travel journalism.[9]

Journalistic travel reports and travel guide literature can be traced back to the tradition of descriptiones and itineraries. Originaly, such travel guides met the demand for written material on the various travel routes of the Crusaders, the Jerusalem pilgrims, and later the young nobles and scholars on their Grand Tour—already showing an service-providing orientation. This continued in the emerging railway and tourism age in the 1830s with John Murray's *handbooks* for travellers and Karl Baedecker's *travel guides* for destinations at home and abroad. The aim of these travel guides was practical—not to offer a new worldview, or redefine the self-understanding of one's own culture or convey a new understanding of the foreign. In addition to practical information to facilitate travel, the *Baedecker*—the name soon became synonymous with travel guides in the German-speaking world—established the *tourist gaze*. They determined what was considered worth seeing and thus channelled curiosity. The curious became canonized—and the view, which had begun expanding to the whole world, with the spectacular and the extraordinary, was narrowed again.[11]

The travel and perception form of tourism, with its developmental history of some two hundred years, thus forms the exact opposite of what serious ethnologists seek to do: foreign cultures are exoticized as counter-worlds, so that they can be exploited as travel destinations, as attraction of the foreign, as aesthetic fascination by the tourism industry. To portray foreign cultures in their normality, rendering them understandable, would contradict the market-economic logic of capitalism twice: first, the tourism industry, which would have to reconceptualize the long-distance tourism market, and second, the culture and media industry, which lives from the marketing of the sensational and extraordinary.

Before the age of film cameras and smartphones, foreign correspondents and long-distance tourism, photographically illustrated travel descriptions were the main supplier of European knowledge about other civilizations.[12] They provided the raw material of data, images, and characters that was further processed by those who stayed at home—by the predominantly male researchers, philosophers, poets, and authors of the penny dreadfuls, as well as by the journalists in the illustrated magazines. They shaped the images of other countries and cultures permanently.

Typical of this are the images of the savages which were created by Karl May, born in 1842, one of the most successful German-language authors of the twentieth century. His German-language edition totalled 1.3 million copies in 1913, one year after his death; by 1970 it had already reached 50 million, not counting the paperback editions. Not the German colonies of the time, but the savannas and prairies of North America were the setting for his descriptions, areas that he had never visited at that time, the years 1860 to 1880. However, thousands of impoverished farmers, workers, and weavers from Silesia and Saxony emigrated to North America during this time to seek their fortune there. These novels of Karl May are fairy tales, stories about a promised land, in which all the characters are invented.[13]

Quite different and yet also belonging to the genre of longing literature are the Himalayan hero epics of Reinhold Messner.[14] Although the ego and the search for meaning of this man, probably the most famous mountaineer of our times, are always central, the mountain dwellers in it do not appear solely as carriers of expedition equipment. After climbing all the world's eight-thousanders, he founded the Messner Mountain Museums in South Tyrol; one in Bruneck in the Puster Valley is entirely dedicated to the mountain peoples of the world. Whereas

Messner explored the peaks, literally travelling to the end of the world to come to his insights, Karl May was a mental traveller. His radius of action was limited to a small part of Germany, and for some years even to a cell of six square meters of a saxonian prison, when he was sentenced by district court because of repeated fraud. This happened before he started his writing career.

The media and the supplying agencies for public relations and marketing are indispensable purveyors of beauty and worthwhile experiences in this business of longing. They rave about retreats in the Alps with wellness Buddhism as a signature treatment, as well as ultimate culinary highlights with Fado accompaniment in Lisbon's Bairro Alto, or picturesque flea markets in the backstreets of London. Songs, films, magazines, and interviews evoke the fascination of New York as the city of cities. Of course, there is no room left to note the daily nightmare of city neurotics, the drug scene, or the underlying violence. Reality takes a vacation. In hardly any lyrical pieces on island dreams or Secret Escapes destinations does the local tourist garbage stink to high heaven, nor are the actual living conditions of the island inhabitants noted, or the unequal benefits of tourism discussed. If you prefer to buy suntan oil instead of heating oil, fly with us to Anywhere, the idyllic warmwater destination! Away from everyday life, onto the plane, happiness is bookable—this is the message proclaimed in millions of travel magazines and the tourism pages of newspaper supplements.

"Tourism is nothing more than the attempt to physically realize the romantic dream projected into the distance", wrote the far-sighted German author Hans Magnus Enzensberger as early as 1958. "The more the bourgeois society became closed, through its values, norms, and became more restrictive, repressive, the more strenuously did the citizenry seek to escape it as tourists."[15] Little has changed to this day—tourism as escape system has matured to a convention, now comprising up to some 15% of the world gross social product. It has become part of the Western lifestyle with the establishment of the industrial-economic social system. About one-sixth of the world's population goes on trips; two-thirds of the population in the Western industrialized countries regularly go on vacation. Freed from the constraints of the industrialized world, a separate culture industry has emerged. The escape from the all-pervading commodity world has itself become a commodity, a highly diversified market offering leisure, recreation, and travel. This fully corresponds to the logic of a capitalist social order, which markets everything—including the longings it produces through its own contradictions.

We can note striking similarities between the global media industry and tourism. Both channel escape movements from the deficit-perceived everyday life; both act as ambulatory therapy rooms, which are visited temporarily. Both provide dreamworld offers. Tourism is seen as an important tool whereby people try to break out of the expanding identity crisis. It is a tool like the ideologization of that borrowed identity, which emphasizes belonging to only one community, one group, one nation—which at the same time means a repression or exclusion of the others.[16] Xenophobia and exoticism are both avoidance strategies. We could say that xenophobia corresponds to an attitude where we avoid the foreign in order *not* to have to question our own. And with exoticism, we go abroad in order not to have to change anything at home.

In the copious travel images and tourism literature, there is only the perfect world—although by now we know that tourism not only generates pleasure and relaxation, but also has the power to destroy cultures and landscapes. Indeed, tourists often destroy what they are looking for by finding it—in large numbers and at the same time. On the other hand, the enormous economic importance of tourism is reflected in thousands of jobs: as a global service industry, it generates incomes, value creation and prosperity in many regions. Tourism has also become a motor of social and cultural change.

The world of images of linear media as well as digital, on the second screen of smartphones, plays an essential role in the production of expectations, role models, stereotypes, and clichés. Like the Austrian and German homeland films of the post-war period, today's catalogues, travel pages, and longing literature deliver beautified realities as well as faked images. If the holiday destinations under the Southern Cross or in the Alps were not advertised as Disneylands inhabited by stereotyped Indians or lederhosen-wearing Alpine Austrians and alphorn-toting Swiss, perhaps those false expectations would not arise, the fulfillment of which makes tourists behave like "cooled soldiers"—as Jean-Paul Sartre once called the invasion of the tourist masses. Long-haul tourism to the countries of the global South often bears a sad resemblance to a neo-colonialist event: no consideration for people, cultures, and ecology in the visited "developing" countries.[17] But also in the overcrowded centres of European mass tourism, the destructive power of tourism is evident—in Amsterdam, Barcelona, Venice, Dubrovnik, Florence, and many other shrines of European civilization.

What the anthropologist Claude Lévi-Strauss noted in his *Tristes Tropiques* (1955), about the pleasure of travelling and travel reports in the face of the progressively irreversible destruction of nature, applies even more today, in our age of nuclear threat, the destruction of ancient cultures, and looming climate disasters: "This is how I understand the passion for travel reports, their madness and their deception. They give us the illusion of something that no longer exists and yet must exist so that we can escape the crushing certainty that twenty thousand years of history have been squandered."[18]

Adventure Lite

Adventure vacation, adventure playground, adventure park, adventure club, adventure magazine—and yes, adventure wet shave. Where standardized routine determines everyday life, even dealing with shaving foam and razor blades becomes an adventure. *Experience marketing* is the media magic formula that turns simple strolls into adventurous, daring feats of trekking. Setting off, being on the move, experiencing new things, the search for risk and danger, the unexpected and unpredictable, combined with geographical change—ventures with uncertain outcomes. As early as 1911, the German sociologist and philosopher Georg Simmel recognized that all features of bourgeois adventure are embedded in a system that everyday life dictates, making adventure an important part of our lives.[19]

The longing for adventure always implies stepping *out* of a cultural order, revolting against restrictions. This applies particularly to sexual adventures, because erotic escapades break with taboos and the bourgeois sexual order, questioning fundamental values. The first South Sea explorers glowingly reported on the carefree life and love forms of the islanders; and people looked enviously to France, whose culture was assumed to allow greater freedom: A Frenchman—a grand seigneur, two Frenchmen—a love couple, three Frenchmen—an ideal married couple. To do it so *à la française* violates the prevailing morality, but is part of the stock of immoral wishful dreams and fantasy worlds of entire generations. Paris is still considered the ultimate world capital of love.

Whereas the first adventure travellers were either young nobles or scientists, today single or childless city dwellers, predominantly males in academic professions and thus intellectual workers, make up the majority of adventure tourists and long-distance travellers. The typical globetrotter is in his prime and can satisfy his desire for exotic holidays thanks to his secure income. While the flower-power children of the 68s generation fled from the compulsory and surplus society into the exotic distance of Indian ashrams to live the simple life, by the 1990s the typical adventure tourist had become modern. The adventure in the form of experience vacation, as an expedition, discovery trip or as adrenaline tourism advertised in catalogues, was made socially acceptable as a distinguishing feature, for demarcation from the masses and as a sign of individuality. Since then, the profile of the adventure travellers has been dominated by fashion-conscious adventurers. With the commercial spread of adventure travel, more and more people are venturing into exotic countries, as adventure vacations have become safer and safer—how paradoxical! With the spread of the Internet, the connection to home base could be maintained in even the most remote regions. Tourism routes today serve as investment guides for telecommunication companies and Internet providers. The WiFi standard is up to date in many underdeveloped but tourist-developed areas—while the sanitary facilities remain at the level of the Middle Ages.

In the age of road-mindedness and unbridled motorization, the experience of the foreign occurs much more frequently en route: in transit from starting point to end point, with only fleeting glimpses of the landscape. The panoramic view perceives only contours: the complexity of a cultural space is reduced to a few striking sights at most. Another popular form of foreign 'experience' is the trip with the mobile home. Here you can be on the move, but in a familiar environment, with your favourite beer and TV programmes, travelling so to speak in your own living room. Just as Asian tourists in Europe tend to appear in battalion strength, consuming the foreign from a safe distance, we can see every group trip from the aspect of risk minimization. Desert trips by jeep in convoy, tourists crossing the Alps in securely roped parties, city tours for retirement home residents in air-conditioned coaches—being in a group offers the feeling of security for dealing with the unknown, foreign and unpredictable.

The Stranger as a Construct

> *V: A stranger is only a stranger in a strange place*
> *K: That is not incorrect. And why does a stranger feel strange in a strange place?*
> *V: Because every stranger who feels strange is a stranger, and remains so until he no longer feels strange—then he is no longer a stranger.*
> *K: And what are locals?*
> *V: The local may not know the stranger, but he recognizes at first glance that he is dealing with a stranger.*
> Karl Valentin, *Monologues and Dialogues with Liesl Karlstadt.*

The foreign is a construct that exists only in relation to the familiar. The characteristics of "the foreign" that appear foreign are perceived as normal in their own environment, but as foreign outside of their context. Foreignness is not a property of people or things, but an attribution that defines distance and difference within social relationships. Not only the foreign is marked, but also the familiar. If the degree of foreignness of people, cultures, landscapes has a significance for tourism—for example because the joy of differences, the curiosity as an emission-free driving force and the contact with other ethnic groups and cultures form an important motive for travel—then we must ask: in what does the attraction of the foreign actually consist?

Essentially, according to Ortfried Schäffter,[20] there are four ordering schemes, modes of experiencing the foreign, that cover the spectrum of possible experiences between fascination and threat. In the first scheme, we interpret foreignness as a sounding board for the familiar and assume a fundamental harmony of differences. The basic fusion typically comes to expression in the wine-soaked Viennese song "Menschen, Menschen, san ma alle—People, people, we are all human beings". Such fraternization across the borders of culture, class, and gender is occasionally interpreted as a unifying component of tourism, which essentially consists of an offer of surprising things, moments of enjoyment—and interculturally inconsequential experiences.

The second scheme, according to Schäffter, sees foreignness as a counter-image, as a negation of familiarity, which leads to the exclusion of the different. The foreign becomes the "natural enemy". The focus is not on the common, but on the contrasting, on the borderlines. This view dominates in the refugee discourse—tourists come and go, but refugees and asylum seekers come and want to stay. When people from foreign cultures immigrate and the majority population wants to delimit and distance itself, conflicts of belonging emerge.

The third scheme interprets foreignness as an opportunity for self-completion and rounding off completion, with the foreign appropriated as a structural complement—African dancing, Arabic cooking, Ayurveda and Feng Shui, Mangas and Mangos, Reggae and Bossa Nova, etc., are integrated into one's own lifestyle. The foreign is seen as a field of learning, where curiosity and risk-taking are prerequisites. We visit those regions and cultures from which we expect a completion of our own personality. Just as the ultimate purpose of any travel activity entails a healthy return, a stimulus, a changed attitude, perhaps even a "different" person, is brought home as a trophy.

Albert Camus is credited with the insight that we are more vulnerable when travelling than in our usual environment. What makes the value of travelling is fear, because the journey into the unknown has less to do with pleasure than with a form of asceticism. Far from our homeland and our language, we are overcome by an indefinite fear, and feel the desire to return to the protection of our old habits. "At this moment we are feverish and at the same time permeable"—travelling as a higher and more serious science that leads us back to ourselves.[21]

In the first three variants of experiencing the foreign, despite the differences, the foreign is incorporated into one's identity. With the second scheme, the demarcation is an identity-forming scheme. The forth scheme, which understands foreignness as complementarity, assumes a fundamental difference and non-appropriability, but through diversity the world becomes round: Just as wave and particle together make up light, the familiar and the foreign are indispensable, mutually dependent components of social existence and reality.[22]

Tourism and the culture industry have become global phenomena that tend to obliterate the separateness and uniqueness of cultures. In cultural theory, there is talk of world citizens without their own territory, of thoroughly hybrid, transcultural personalities. Although elements of each culture tend to form influencing factors for all other cultures, practice shows that cultural differences are often difficult to cope with. Misunderstanding and intercultural conflicts, not successful communication between nations, are often the result. Encountering people from another culture can evoke deep feelings of helplessness, fear, and aggression in tourists. In most cases, closer contact is avoided or reduced to the bare minimum; but the holiday situation also encourages recklessness and curiosity, with the lure of playing with the thrill of fear.

Cultural confusion or even culture shock[23] arises when our own ideas about the correct interpretation of the world no longer hold true, when cultural otherness can no longer be meaningfully integrated into our own system of experience. Food, pests, noises, smells, different standards of hygiene, traditions, gestures, language, unfamiliar understandings of proximity and distance—all this can generate stress and fear, provoke emotional imbalance and result in disorientation. Shocked in this way, we may long to return home, or remain in the hotel, avoiding contact with strangers, perhaps drowning our fears in alcohol. Others get irritated at petty details, feel helpless, constantly washing their hands, perhaps even refusing to eat. An attempt to reduce stress ends in flight, in fight, in disgust and rejection. Prejudices can easily mutate into enemy images, become xenophobic attitudes and lead to racist utterances. The disappointment of illusion results in the downgrading of the locals to *underdeveloped*.

By contrast, the optimistic individual tries to use humour and tolerance to come to an inner acceptance of the circumstances. Ultimately some advantages of the local culture are perceived, and one tries to engage with the given circumstances. In this way, one not only lifts oneself out of depression, but ultimately also successfully establishes intercultural contacts. In the extreme variant, the outsider falls in love with the foreign culture, dresses and behaves like locals, adopts the lifestyle of the foreign culture and for a short time becomes almost an indigenous person himself.

If one engages so intensively with the foreign, the perfection of the adventure is not that it begins and ends in the span of one night, as the Italien novelist Italo Calvino claims in the story *Journey of an Employee*. It is probably the imagine of the Noble Savage, the solid projection of the dream of a simple life, the antithesis to European civilization madness and its constraints, that make elements or models of other life so attractive for tourists—at least for the duration of a holiday stay. The seemingly casual workday, the simplicity, being poor but happy, the carefree joy of existence, social equality without striving for possession or the rhythm of life determined by nature—all these contrast with modern Western culture.

In particular, the relaxed handling of time, the significantly slower pace of everyday life, fascinates Europeans stuck in tight time-corsets. In this image of the perfect world—whether on the cloudless Peruvian Altiplano or the mild African summer night, on the palm beach or in the luxury tent of a camping resort in the Serengeti—any reference to reality is consistently missing. Tourists are blind to the politically ugly or the hardships of life in alpine regions and in tropical latitudes. Few want to hear or see anything about hunger, poverty and similar unpleasantness. Their fantasies are based on the images circulated by the glossy products of the tourism and entertainment industry. Only well-prepared and experienced travellers break out of this pattern: they can critically view their environment politically, ecologically and culturally, and identify significant deficits or deviations from the promise of utility value.

The Foreign Looks Stranger Than It is

That the representations of foreign cultures and countries are a media or tourist construction should not surprise us in the age of technological compression and relativization of space and time. The mass media generate their chosen images of the world and circulate them en masse, trade in stereotypes that massively influence and determine our thinking and feeling. Few regions have become as mystified and alienated by illusions as Tibet. Chakra meditation for male and female sexuality, yak butter tea for endurance, blessings from the Dalai Lama for a fulfilled life—everything is stuffed into the construction of Tibet. Tourists search in this spiritually charged destination of longing, which is constantly fed by Western literati and Hollywood images, for the *place of eternal happiness*. Because distant protests against Chinese oppression do little to help the nomads and small farmers on the roof of the world, H.H. the Dalai Lama is reverently courted by the international media public, his story tailormade for the media. A veritable living god enters the world stage. Morally impeccable and with great charisma, he spreads the message of peace, to the distress of the bureaucrats in Beijing—this simple monk, as he calls himself, reaps applause like a rock star and becomes a global superstar.

From the mysticism of ancient Tibet, the media have conjured up a dream world—as meticulously demonstrated by Tibetologist Martin Brauen[24]—and the growing tourism industry is doing its best to revive the legends. The uniqueness of a trip to the roof of the world lies not only in being more than 4000 m above sea level, but also in the height of the price. Tour operators seize the images spread by the media and the connection between imagination and dream; fiction and tourism and expressed in the dream journey or in dreamland Tibet. Realities are replaced by emotional elements. While in practical life these projection processes are constantly limited, this is not the case in film or tourism, where one can dream without consequences. If China bans access to Tibet for Western tourists, as is often the case, all that remains of such a trip is indeed an illusion—the hope that the barriers will soon open again and that the Tibetans will retain a quantum of self-determination.

Not only in the Himalayas are tourists in search of the place of eternal happiness or fulfillment of the promises that films and literature have created. Even if in film, as Paul Valéry thought, all attributes of the dream are equipped with the precision of the real, tourists still shape their perception of reality themselves. They travel to the images they already know from films, postcards, and postings, and then take that photo with their own cameras. The experience consists in the exaggeration of this one moment: the rest may sink into the shadow of uncertain memory. Nothing is more deceitful than one's own memory when it comes to remembering. But the camera captures photographically what seems necessary for self-assurance. The foreign has been perceived by one's own eyes and the encounter documented. The exotic souvenir in its trivialized form as Airport Art is part of a successful long-distance journey like the photo evidence. Recognition at home creates additional satisfaction, making the venture *ex post* a special kind of experience, despite the trials and tribulations involved.

European fantasies have been reducing the exotic worlds to stereotypes[25] for centuries. Tourism, fashion, the entertainment industry domesticate the foreign—exoticism is everywhere today—while preserving it as a surface for projecting their longings and dreams. This highly contradictory and divided attitude has always characterized the West in its reception of the Other.[26] European painters designed oriental scenes that had nothing to do with reality: they painted their innermost wishes and dreams from their souls. While beach holidays on the Turkish or Croatian Mediterranean coast currently top the popularity scale, having, say, a Turkish woman or Croatian man as neighbors at home is far less appealing; similary with the refugees from African countries or from the war zones of

Asia Minor, the cradle of humanity. Our attitude towards foreign cultures has always been ambivalent and is still characterized by self-interest today.

Dreams turn into traumas when dreams are violated or expectations remain unfulfilled. The exotic world is not a dream world, not an escape hatch, not an idyll—but it is marketed as such, with the underlying motif of longing. In truth, it consists of many distant endangered worlds that we bring closer with our fantasies. Tourism simplifies the arduous approach, the kitsch industry trivializes and cheapens everything: we live in a cliché world of stereotypes, drowning in masses of exotic images.

Deeper, experiential travel satisfaction is more likely to arise in the tourist relationship between the familiar and the foreign when we have prepared well for the journey, adapting our expectations to the setting, without destroying anything on site. Today's tourists are more seekers than finders. Especially the rich and mobile elites, constantly on the move and without visas or residence permits, convey the impression of territorial homelessness, always in search of novelties to consume and incorporate. The hunt for dream worlds does not allow lasting bonds—only distraction. Often it remains with the tourist gaze, the astonished look and gawking. The postcolonial nursery allows no more, or lacks tools such as respect, humility, empathy and openness, or decoding and interpretation techniques. But that should not deter anyone from daring to dance with the foreign, the exotic, to take a first step, however timid.

Endnotes

1. Mario Erdheim, Zur Ethnopsychoanalyse von Exotismus und Xenophobie (On the ethnopsychoanalysis of exoticism and xenophobia), in: Mario Erdheim (ed.),

Psychoanalysis and the Unconscious in Culture, Frankfurt: Suhrkamp 1988, 258–265.
2. Julia Kristeva, *Fremde sind wir uns selbst* (We are strangers to ourselves), Frankfurt: Fischer 1990.
3. The history of European–overseas encounters is a deeply contradictory one, as Urs Bittlerli shows in his *Die 'Wilden' und die 'Zivilisierten'* (The 'Wild' and the 'Civilized'), Beck: Munich 1991. The colonial empires paved the way for the current North–South conflict through their conquests and centuries of rule. See Jean Ziegler, *Die neuen Herrscher der Welt und ihre globalen Widersacher* (The New Rulers of the World and their Global Adversaries), Munich: Goldmann 2003.
4. For a detailed discussion of the role of the mass media see Kurt Luger, Perfekte Völkermissverständigung (Perfect Misunderstanding of Peoples), *Journal für Entwicklungspolitik (Journal for Development Policy)*, 3/1990, pp. 5–23.
5. Clifford Geertz, *Dichte Beschreibung* (Thick Description. Contributions to Understanding Cultural Systems), Frankfurt: Suhrkamp 1991.
6. Peter Matthiesen, *Auf der Spur des Schneeleoparden* (On the Trail of the Snow Leopard), Berne: Matthes & Seitz 1978.
7. Mircea Eliade, *Indisches Tagebuch. Reisenotizen*, Munich: Diederichs 1996, here 219.
8. A good introduction to the genre of travel literature is provided by Peter Brenner, *Der Reisebericht (The Travel Report)*, Frankfurt: Suhrkamp 1989.
9. For a detailed discussion, see Daniel Cuonz, Heine's Unrast, Poetologie einer Selbstverortung (Heine's Restlessness, Poetics of Self-Location), in: *Zeitschrift für Literaturwissenschaft, Ästhetik und Kulturwissenschaften* 2/2018, 165–184.
10. For a Comprehensive Presentation of Travel in this era, see Justin Stagl, *Eine Geschichte Der Neugier (A History of Curiosity, The Art of Travel 1550–1800)*, Vienna: Böhlau 2002.

11. For a discussion of what is referred to as the "tourist gaze" in Anglo-American literature, see John Urry, *The Tourist Gaze*, London: Sage, 1990, and Dean MacCannell, *Empty Meeting Grounds, The Tourist Papers*, London: Routledge 1992.
12. Kurt Kaindl, *Harald P. Lechenperg. Pionier des Fotojournalismus* (Pioneer of Photojournalism 1929–1937), Salzburg: Otto Müller 1990.
13. Helmut Schmiedt, *Karl May, Leben, Werk und Wirkung (Life, Work and Impact)*, Frankfurt: Beck 1992.
14. Dominik Siegrist, *Sehnsucht Himalaya* (Longing Himalaya, Everyday Geography and Nature Discourse in German-speaking Mountaineer Travel Reports), Zurich: Chronus 1996.
15. Hans Magnus Enzensberger, Eine Theorie des Tourismus (A Theory of Tourism), in: Hans Magnus Enzensberger, *Einzelheiten I, Bewußtseins-Industrie* (Details I, Consciousness Industry), Frankfurt: Suhrkamp 1967, 4th edition, 179–205, here 190–191.
16. Kenneth Gergen, *The Saturated Self, Dilemmas of Identity in Contemporary Life*. New York: Basic Books 1991.
17. For a detailed discussion on this connection, see Herbert Baumhackl, Gabriele Habinger, Franz Kolland, & Kurt Luger (eds.), *Tourismus in der "Dritten Welt"* (Tourism in the "Third World", Discussing a Development Perspective), Vienna: Promedia 2006.
18. Claude Lévi-Strauss, *Traurige Tropen (Tristes Tropiques)*, Frankfurt: Suhrkamp 1978, here 31.
19. Georg Simmel, Das Abenteuer (The Adventure), in: Georg Simmel, *Philosophische Kultur* (Philosophical Culture. Collected Essays), Berlin: Wagenbach 1983, 13–26.
20. Modi des Fremderlebens. (Modes of experiencing the foreign. Interpretation patterns in dealing with foreignness), in: Ortfried Schäffter (ed.), *Das Fremde* (The Foreign—Experience possibilities between fascination and threat), Opladen: Springer 1998, 11–44.
21. Albert Camus, *Tagebücher* (Diaries 1935–1951), Reinbek: Rowohlt 1972.

22. See Erich Hamberger, *Kommunikation und Komplementarität* (Communication and Complementarity—Fragments of a Transdisciplinary and Transcultural Communication Theory), and Thomas Herdin & Kurt Luger, Kultur als Medium der Kommunikation (Culture as a Medium of Communication), both in: Erich Hamberger & Kurt Luger (eds.), *Transdisziplinäre Kommunikation* (Transdisciplinary Communication), Vienna: Österreichischer Kunst- und Kulturverlag 2008.
23. *Culture shock* is a central theme of intercultural communication and is richly illustrated by curve models. It refers to the fall from euphoria into a feeling of being out of place, and then slowly adapting to the foreign environment and developing understanding and competence. For expatriates, this is a central problem; in tourism it is more often referred to as 'confusion'. See Petri Hottola, Culture Confusion: Intercultural Adaptation in Tourism, *Annals of Tourism Research,* 2/2004, online: https://doi.org/10.1016/j.annals.2004.01.003.

    The book market reacts to such irritations with a wealth of travel know-how literature. A basic discussion is provided by Thomas Herdin & Kurt Luger, *Wir und die Anderen* (We and the Others. Intercultural Encounter Field Tourism), in: Roman Egger & Thomas Herdin (eds.), *Tourismus im Spannungsfeld von Polaritäten (Tourism in the Field of Polarities)*, Vienna: LIT 2010, 337–357.
24. Martin Brauen, *Traumwelt Tibet (Dream World Tibet)*, Berne: Paul Haupt 2000.
25. Martina Thiele, *Medien und Stereotype (Media and Stereotypes, Contours of a Research Field)*, Bielefeld: transcript 2015.
26. See Rolf Neuhaus, *Reisen nach Ophir (Travels to Ophir: On the Search for Happiness in the Distance, From Humboldt to Hesse, from Timbuktu to Tahiti)*, Wiesbaden: Marix 2020.

3

Places of Happiness, Mobile Privatization and Emotional Geography

Knowest thou where the lemon blossom grows,
In foliage dark the orange golden glows,
A gentle breeze blows from the azure sky,
Still stands the myrtle, and the laurel, high?
Dost know it well?
'Tis there! 'Tis there
Would I with thee, oh my beloved, fare.
Johann Wolfgang von Goethe, Wilhelm Meisters Lehrjahre, 1795 (translated by Walter Meyer, quoted from https://www.lieder.net/lieder/get_text.html?TextId=6461.)

Come a little with me to Italy,
come with me to the blue sea,
and we 'll pretend as if life
is a beautiful journey.
Sung by Catharina Valente & Silvio Francesco.
Text Kurt Feltz, Music Heinz Gietz, 1956.

An essential characteristic of our Western industrialized culture is mobility—or rather, total mobilization, the seemingly complete control over space and time. In 24 h or less, we—or the financially privileged in today's society, the *kinetic avant-garde*—can fly to almost any place in the world, and young Interrailers can get to know three European train stations in the same period of time. At the push of a button, we have live access to the images that are made accessible to us at home and can be switched on and off at will or swiped away on the smartphone. Two achievements are responsible for this globalisation of culture and communication: international tourism, and the expansion of the media and culture industry, which, in conjunction with new technologies like the Internet, has created a globally available image market at any time.

The spread of industrialization, with urbanization and greater regional mobility, with increases in income and the saturation of everyday consumer needs, accompanied by a flood of reasonably priced tourism offers, the diversion of societal differentiation and prestige desires—all these have led to mass tourism in the Western industrialized countries. This applies to the many popular and thus over-visited places—from historical old towns, to ski slopes in the Alps and beach-strewn sea coasts, which fell victim to mass tourism, but which also provoked and sought it, as instrinic elements of the lifesyle of industrial society. Depending on the level of prosperity, holiday trips belong to the periodic routine that brings variety to everyday life. They increase the enjoyment concentration of an experience society that understands tourism as mobile leisure.

How the media and tourism industry can bring about cultural change is illustrated by a look at the 1960s, when vacation and tourism became culturally anchored as part of the lifestyle. This was a decade of boundary blurring, as regards societal values and moral concepts, but also in the satisfaction of dreams. In that decade, mobile privatization noticeably shaped the lifestyles of many Europeans for the first time.

This came to mean the search for individual, private freedom and thus identity. It found expression mainly in consumption, and in the design of our living quarters and in the family area. In Austria, television became the main technical achievement of the 1960s, comparable to the refrigerator in the previous decade. With rising incomes, mass production and cheaper device and gadgets, coupled with demands for convenience and prestige purchases, the number of television households increased from 200,000 in 1960 to 1.4 million in 1970. For many people, this privatization was also accompanied by withdrawal from active political involvement.

Moreover, technology now offered the possibility of unprecedented mental mobility and chances to participate in events worldwide. People could expand their horizons, experiencing media events ranging from space flights and moon landings to Olympic Games—and they could directly follow how the USA bombed little Vietnam back into the Stone Age. Television fused the horrors of wars and mass murders with the normality of living room everyday life.[2]

During the same period, the number of cars in Austria increased from 404,000 to 1.2 million, the minimum vacation time was increased from two to three weeks, and the weekly working hours were gradually reduced to 40 h per week according to collective agreement. This increase in available time favored the physical form of mobility and led to a collective departure for vacations abroad. Holidays on the Upper Adriatic beaches became the epitome of foreign travel for Austrians and Germans alike, now affordable for the less affluent. Like the gems of the Austrian province, the dream world of trivial novels, films, and hit songs with their mixture of sea, music, and romantic love became an easily usable landscape backdrop. Warmth and the South formed the escape spaces from the "cold world" often lamented in hit songs. The myth of the South Seas also found offshoots in schmaltzy songs with Hawaiian guitars and Aloha choruses: indeed, dreams of a South Seas cruise still top the wish list of many would-be travellers.

More and more people could disconnect for a few weeks of holiday. The individual radius of action was significantly expanded, and the increasingly motorized traffic developed into mass tourism, which Austrians increasingly experienced as guests, but even more so as those being visited. In 1954, only 42% of foreign visitors came to Austria by car; by 1960 it was already 84%, and little has changed to this day. New means of communication, private automobiles and highways, air traffic and television increasingly diminished the hierarchy between near and far. At the same time, they promoted the formation of identities that sought to absorb the foreign into the familiar. The desire for boundarylessness became an industry of its own.[3]

The expression of individuality through the spatial dissolution of boundaries, with air travel and steadily expanding private car traffic, has contributed significantly to the negative impact on the climate in the form of greenhouse gas and accelerated climate change. According to the UN World Tourism Organization, more than five percent of all human-caused CO_2 emissions are attributable to the tourism sector—and the trend is rising.[4]

The Perception of Space

Consumer attitudes towards nature—which is a primary goal of tourist travel—is an expression of a high-speed society that is also characterized in tourism by the rapid use of the experiential, the rapid change of attractions, and the restless rush to new things, new experiences. In the process, quality is lost due to speed, because perception inevitably becomes superficial or fleeting, and understanding of the context often cannot arise at all.

Landscape does not exist without an observer. What we understand today by the term *landscape* originated in the eighteenth century. Before that, nature or the environment as perceived by humans was seen as creation, an ordered whole that did not require any special aesthetic mediation or interpretation; the environment was in a practical or moral relationship with humans. Only when nature became the object of scientific research, technical use, and economic appropriation, broken down into its components, did it become necessary to re-assemble it into an aesthetic whole. The material of nature was transformed into the *structure of landscape,* as Georg Simmel wrote at the end of the nineteenth century in his *Philosophy of Landscape.*[5]

Whoever looks into the landscape—the term was coined by Dutch painters, to denote the central perspective representation of a beautiful area, images of rural scenes—chooses a location and, starting from this, a section, a part of the whole. The sensitive observer wishes to experience wholeness in this section. It became the task of artists and poets to merge the individual parts into a whole, into a composition. Their aestheticization of certain landscape sections, viewpoints, and fields of view, is a prerequisite for tourist landscape enjoyment. The term landscape entered everyday language and now, within the framework of culturally shaped perception patterns, denotes the result of aesthetic-subjective perception, in which a sensitive observer views an area shaped by nature alone (natural landscape) or by nature aided by human intervention (cultural landscape) as a harmonious, individual, pictorial whole. Thus, the unity of a landscape or its beauty results not from a causal connection of the objective objects in an area, but from the aesthetic, selective, and synthesizing perception. Select and assemble appropriately from a subjective perspective—that is how the landscape image is created in the eye of the observer.

Within the context of enlightenment and criticism of civilization, landscapes were no longer interpreted as subjectively aesthetic wholes, but as objectively given regional units, as a unique organic connection between land and people. This was understood as the result of successful cultural development, aesthetically expressed the beauty of the landscape. Since then, in European culture, the viewing of landscapes—especially small-scale pre-industrial cultural landscapes—symbolizes the ideal of harmonious, sustainable, unique regional human-nature units or socio-ecological systems that need to be protected against globalization and industrialization, against concretization and urban sprawl.[6]

Technical interventions in nature through the construction of settlements, commercial areas, roads, parking spaces, roundabouts, bridges, cable cars, power poles, hydropower plants, water reservoirs, river regulations and similar changes to a previously untouched landscape can act as massive disruptive factors and provoke societal protest. And then, already in the next generation, they may be taken for granted. What is perceived as untouched, beautiful or destructive changes over the years, becoming the new norm of everyday life.[7]

Tourism is often accused of consuming natural landscapes because it prepares it for 'experience space management'—think of all the ski slopes and reservoirs for artificial snow production, or the construction of hotel complexes, chalet villages, golf courses, etc. on the Mediteranean islands, where exuberant holidaymakers party through the night, and many silver-agers dream of spending the autumn of their lives basking in the sun. On the other hand, we should acknowledge that the restoration of historical buildings and their subsequent tourist use can also make a significant contribution to the preservation of world architectural heritage. The construction of a motorable road or a cable car enables more and more people to experience the beauty of nature or to enjoy the new dimension of a designed landscape as a spectacular sensory experience. Previously, such massive interventions in nature were undertaken purely for economic reasons, and justified with the expectation of high regional value creation. Today, with the threat of advanced climate change, such large projects can be justified only if their overall societal benefit and their environmental compatibility are proven. However, this does not mean that dubious investor speculations can always be prevented!

When Reading Your Postcard, I Hear the Surf Roaring

When we speak today of imaginary or emotional geography, we assign a high symbolic as well as emotional significance to landscapes, regions, cities—specific places within a geographical space. Geographical worldviews are based on evaluations, on ideological guidelines of a political as well as religious nature, on literary representations, photographs, films, dreams and fantasies, which can become symbolic spaces, psychological space constructions or even utopias. This affects our own environment—exemplified, for example, in the ideologically charged concept of *Heimat* ('homeland')—or the distant space imagined as a dream destination. Imagined pictures in the mind make spaces into places, imagined landscapes shape the subjective experience of landscape even before we get to know it from our own observation, because they control our perceptions, desires and ideas of beauty.[8]

Tourists travel to images or ideas. They seek the sensory experience of imagined worlds and create their own experience spaces with imagination and projection. In journeying, tourists seek confirmation of their imaginary geographies, the imagined pictures; their perceptions will largely follow or adapt to these imaginings. Any realities that might contradict the imagination, realities which cannot be reconciled with the expectations and wishes, are kept at a distance. The worldviews triggered by travel literature, photographs and film create a constant continuum of perception, reinforced in the constructions of tourism marketing, which produces illusions and dream worlds. Such worldviews imagine landscapes, their inhabitants and ideal concepts, equipping space with meaning and significance. The mental space, shaped by signs and symbols, merges with the physical space. The imaginary geography semiotizes the space, assigning signs and meaning.

In all societies and among all people, the imagination seeks spaces away from the world of everyday tasks. Festivals and games, rites and rituals, fairy tales and myths feed the imagination—as daydreams and especially travel. Lying between reality and imagination, travel allows us to take a break from our standardized everyday lives and wearisome routines. It doesn't matter whether this is done out of aversion to it or because we seek amusement, hoping to enrich our lives through new experiences. No entertainment techniques combines real activity and fictional experience to this extent like travel. The holidaymaker is physically on the move and at the same time enters spaces of the imagination. Tourism combines imaginative activity and physical actions, it finds expression in the realm of imagination as well as in the physical world: it leads into real, tangible worlds, and yet remains attached to the imaginary, to dreams and wishes. In his book *Reiselust,* Christoph Hennig also sees this as a central element that explains the fascination of travel.

Speed and Perception

The quality or intensity of our perceptions or experiences depends on the manner and speed at which we move through space. Landscape does not exist in and of itself, but is created as an image in the minds of its observers. To find a coherent whole from the abundance of coexisting things and impressions thereof, to see a landscape in it, is the cultural achievement of the observer, a creative act of the brain—so argues Lucius Burckhardt, the inventor of the science of strolling, of *Promenadology*.[9] The faster the mode of transport, the more fleeting and coarse-grained the overall impression becomes: the special or typical of a visited area as a landscape can no longer be identified. Landscape paintings shape views and perceptions of beauty; landscape gardens often reveal themselves to the observer through viewpoints that can be reached only on foot. Thus, in today's accelerated society, there is a longing for supposedly intact landscape images—as evident in tourism brochures and in lifestyle magazines, but in reality are accessible only if we distance ourselves from the pervasive principle of mobility.

According to Marc Augé,[10] natural spaces—as imagined Arcadia and ideal refuges of contemplation—are contrasted by a multitude of *non-places* such as airports, train stations, motorway service stations, hotel chains and supermarkets, which are not anthropological places themselves. We cannot feel at home in these places: they are visited in motion in the global network of transport and mobile infrastructures. And as these non-places are constantly increasing and furnishing our everyday experience, the number of places where human society has less influence is shrinking. In the German population, for example, there is a great consensus that natural areas in their still existing originality must be protected from society and subject to usage restrictions, but should remain accessible. In the Alpine countries, you will find today only very small such areas with a wilderness character within the core zones of national parks or protected areas.

The history of the landscape is thus also one of the means of transport, because its experience is a result of the speed at which a space is traversed. Our everyday actions and lifestyle have become increasingly mobile and faster. The journey by train in the nineteenth century—by today's standards incredibly slow—created a two-dimensional panorama world consisting of space and time.[11] Not the direction of one's own movement was to be experienced from the compartment carriage window—only a rapid sequence of views. The train stages a new landscape and the new form of perception of a flowing space–time panorama through speed. The movement of the train through the landscape creates the impression of a moving or changing landscape, its speed makes objects and scenes appear in an immediate sequence. The *panoramic view* from the compartment window requires a quick synthesis by the eye to establish the relational structure. It captures a scenery created by movement whose fleetingness makes the capture of its entirety almost impossible.

Panorama art experienced its heyday around the middle of the nineteenth century. Painters produced huge circular images entailing the detailed reconstruction of the depicted moment, because it is this moment that fixes the point in time of the view. Large panoramas were intended for a broad audience and thus early mass media. The Irish miniature painter Robert Barker, who is considered the inventor of this genre and the terminus technicus (*pan orama—see everything*), took care to have both patented.

Particularly popular were the *Moving Panoramas*, horizontally rolled strip-like canvas paintings. They are considered precursors of film and cinema movies, because they gave the impression of experiencing a passing landscape.

Panoramas of Sydney, Hobart Town in Tasmania or Gibraltar delighted the imperial consciousness of urban visitors in England; on the continent, mountains were the preferred panoramic theme. The summit world of the Swiss Alps became a crowd-puller. Dioramas managed to convey the illusion of the day's course from sunrise to sunset. The greatest audience success was the Moving Panorama, which depicted the ascent of Mont Blanc. The show, which opened in 1852, experienced 2000 performances; its success contributed significantly to the enthusiasm for the Alps and to the promotion of tourism in Chamonix. The cultural philosopher and leading art critic of Victorian England, John Ruskin, noted angrily that this town, located at the foot of the highest mountain in the Alps, was as populated by English mobs as was Piccadilly during the rush hour.[12]

Panoramas allow viewers a 360-degree view of a painting from a central perspective. They stand in the centre, presenting the scenes as if from a tower. Enormous circular images, such as the Sattler Panorama of the city of Salzburg and the Landscape Garden of its surroundings, which measures 5 m in height and 26 m in length and was painted in the late 1820s, appealed to broad audience.[13] The artist Johann Michael Sattler and his family dismantled and packed the panorama and the wooden pavilion necessary for the exhibition and took it on tour, travelling on a houseboat and with wagons for more than ten years through art-loving Europe, where they presented the circular image for an admission fee. Sattler earned a living for his family while also making the beauty of Salzburg known everywhere.

This deployment in Europe, where travel was not yet easy, stands as an early form of tourism marketing or *Location Placement*. It found a continuation about a hundred years later through the Hollywood film *The Sound of Music*, featuring a fairy-tale plot with song interludes, showing enchanting locations in the city and countryside of Salzburg. The filmed version of the musical still attracts hundreds of thousands of tourists, especially from the USA and Asia, to the Salzburg region, offering the radiating and attracting power of images that promise a *Heterotopia*, a place of happiness as a localizable utopia. *Movie-induced tourism*—as this phenomenon ia called in communication science—is clearly a success story of the global interacting entertainment industry.

The Sublime and Natural Beauty

Landscape became the major theme of art around 1800. It appears as a special form of natural space, as the face of the country, directly affecting the viewer in the mirror of subjective feelings and aesthetic interpretation patterns. After the lush and expansive space constructions of the Baroque, the landscape section narrowed: the view shifted from the distance to the near and detailed. Focal points of romantic emotional art fixate the wilderness, where it appears harmonious and idyllic. The age of Romanticism is characterized by a sacralization of landscape, it becomes the object of almost religious devotion. Painting, music, and literature create new mood spaces as remedies against the diseases of civilization, which are essentially attributed to urban areas. The age of discovery of high mountains and seascapes arrives. In the first half of the nineteenth century, driven by romantic currents, enthusiasm for picturesque mountain landscapes emerged—and has remained decisive as a *voyage pittoresque* well into modern times, also stimulating alpine tourism.[14]

As sublime and naturally beautiful, landscape enters romantic-idealist philosophy and literature in the nineteenth century.[15] The concept of the Sublime or Elevated expresses the ambivalent feeling of pleasure and terror in the face of the overwhelming and alien nature, in contrast to sheer pleasure in beautiful works of art. Take the example of the Alps. Until the eighteenth century, the mountains were considered places of horror and a hindrance to movement, and were largely avoided. The same applied to forests, seas and deserts, which were also avoided as hostile places that could be mastered only with great competence and divine assistance. It was through the imaginations of poets, painters, and philosophers that the Alps were to become a longing-laden space of imagination.

In the idea of the sublime, the tamed and aestheticized horror, lie the preconditions for the landscape cult of recent centuries. This set the course for modern nature and mountain tourism, for experiencing the sublimeness of nature or mountain world through one's own eyes as a central travel motive.

This positive image of the Alps is based on an idealized view from the perspective of the lowlands, defining the mountains as a peripheral space, at the same time a retreat and a guarantor for a perfect world, in contrast to urban life and civilization. By ignoring the harsh living conditions of Alpine dwellers, or idealizing them (the noble savage in the figure of the happy alpine farmer) and the interpretation of the landscape as a perfect world from a tourist perspective, came the starting point for today's view of an almost utopian counter-space. From the terrifying mountain world, a postcard idyll was created—at least in the minds of alpine travellers.[16]

Within two centuries, the Alps underwent a revaluation on mental maps. This was developed and practised by an intellectual elite—first in the change of meaning to a place of longing, and then through alpine travels and the emerging alpinism. Rural or alpine space underwent a new coding in the ideas of the dominant urban cultural consciousness. The emotional exaltation of the Alpine phenomenon in bourgeois society bore traces of a natural theology. At the summit, man manifests himself as the master of nature. Previously only God had had the overview: now man had also gained the divine view.[17]

This process was triggered and accompanied by a scientific-rational exploration and domination of space, as well as an aesthetic-emotional turn. Both can also be traced in the context of historical circumstances: massive social and economic changes, a world in rapid transition from monarchical or feudal to republican structures, and the rise of the wealthy urban bourgeoisie, who claimed a political leadership position thanks to their economic power.

This romantic image of the Alps has remained powerful, because it combines the historically original with the expression of an apparently perfect harmony of landscape and man, of colour and 'the air of paradise'. It still features the ancient motif of the ideal landscape *Arcadia*, praised in early pastoral poetry as a place of carefree happiness. We find it today in the advertising messages of tourism marketing as well as on the cover pages of glossy coffee-table books, stereotypical images of a supposedly intact world. This is a primal landscape, presented as authentic and ecologically flawless. Some 250 years after Jean-Jaques Rousseau's rapturous epistolary novel *Julie, ou la nouvelle Héloise*, or the poem 'The Alps' by the Bernese naturalist

Albrecht von Haller, the desire for boundless freedom is still modelled with symbolic images, and the 'untouched' alpine landscapes are advertised as the ideal refuge.[18]

The two central aesthetic categories of this image are the Beautiful and the Sublime. Whoever judges the beauty of an object, according to the German philosopher Immanuel Kant, also asserts a verdict that others should agree with. For Kant, beauty entails subjective universality, about which one may have different opinions, in contrast to the Good or Pleasant, where personal interest in the object plays a role, he defines beauty as "disinterested satisfaction". Agreement is easier to establish in the second category, the Sublime. It has had a greater impact on the recoding and change of the image of the Alps, because it expresses "the absolutely great" of the natural spectacle, or, as the Renaissance scholar Francesco Petrarca put it when he processed his ascent of Mont Ventoux in writing, something that "exceeds every measure of the senses".[19]

Places of Happiness—Heterotopia

A place projected in this way into the near or far distance can be described as a *Heterotopia*. This terminology has found its way into cultural and tourism theory, where it is interpreted as an *Other-Place* or as a localizable Utopia. As counter-sites, such spaces serve the emotional stabilization of societal processes. Michel Foucault,[20] who enriched the spatial-theoretical debate with this term, mentions the garden as the oldest example of a heterotopia. Such a place, filled with greenery, flowing water and surrounded by bushes, is referred to as *Paradise* (literally: a garden) in ancient Persian poetry. Garden motifs are presented in abstract form in carpets, as elements of an ideal place of longing without emotional clouding, friction or rejection.

This contrasts with other emotionally charged heterotopic places such as the cemetery, the prison or the psychiatric institution, which associated with problematic exceptional and special conditions.

The extraordinariness of heterotopic places characterizes tourism spaces as emotional escape rooms, niches in which we can even express and cultivate feelings that violate societal conventions. Heterotopias are quasi-artificial places where everything appears that excludes the everyday. Klaus Kufeld sees in the cruise ship the perfect representation of the quasi-utopian integral with good food, bar music and deck chair, and wishful thinking and the good life have a temporary place—all this on the open sea and and only the blue sky stretches over the soul dangling self! The cruise ship heterotopia is for him a comfortable home on travels.[21]

The holiday trip in a car with a caravan trailer or with a camper van can be said to be a less-luxurious variant that also enables heterotopic experiences in the context of one's own family and domesticity.

Karlheinz Wöhler[22] refers to tourism habitats as spaces of desire and as *Places of Happiness*. They are staged with experiential elements and imbued with meanings, sensually explored and emotionally appropriated by tourists: an external happiness, about earthly paradises, filled with extraordinary highlights, which are propagated by the media as ideal images. Capturing such images of the extraordinary and moments of bliss has become the task of photography—once the domain of travel photographers and photojournalists. Today, a digital camera is standard equipment for every smartphone, giving us even greater opportunities for preserving the fleeting, transient and beautiful. Photography implies a projective moment—the anticipation of a later desired memory, the capture of a mood that we wish to experience again and again.

Endnotes

1. Raymond Williams, *Television—Technology and Cultural Form*, Glasgow: Routledge 1973.
2. Die Ferne der Nähe und die Nähe der Ferne (The distance of the near and the near of the far. TV images and comments on everyday life of the 60s), in: Holtfreter Jürgen et al. (eds.), *CheSchahShit The sixties between cocktail and Molotov*, Berlin: Elefanten Press 1984, 50–61.
3. Kurt Luger & Franz Rest, Mobile Privatisierung (Mobile Privatization. Culture and Tourism in the Second Republic), in: Reinhard Sieder, Heinz Steinert and Emmerich Talos (eds.), *Österreich 1945–1995* (Austria 1945–1995), Vienna: Verlag für Gesellschaftskritik 1995, 655–670.
4. https://www.e-unwto.org/doi/book/https://doi.org/10.18111/9789284416660, December 2019, accessed 30.07.2021.
5. https://socio.ch/sim/verschiedenes/1913/landschaft.htm, 30.07.2021.
6. See Thomas Kirchhoff, 2012, *Landschaft* (Landscape: Basic concepts of natural philosophy). http://www.naturphilosophie.org/landschaft, 21.05.2021; Ian Thompson (ed.) *Rethinking Landscape, A Critical Reader*, London: Routledge 2008.
7. Martin Burckhardt, *Metamorphosen von Raum und Zeit* (Metamorphoses of Space and Time—a History of Perception), Frankfurt: Campus 1997.
8. On theory and semiotics of space and spatiality see Stephan Günzel, *Raum—Eine kulturwissenschaftliche Einführung (Space—An introduction to cultural studies)*, Bielefeld: transcript 2017; Michael Seebacher, *Raumkonstruktion in der Geographie* (Construction of Space in Geography), Vol. 14, Abhandlungen zur Geographie und Regionalforschung, University of Vienna 2012, online: https://unipub.uni-graz.at/obvugroa/content/titleinfo/378523/full.pdf; Karlheinz Wöhler, Andreas Pott & Vera Denzer (eds.),

Tourismusräume. Zur soziokulturellen Konstruktion eines globalen Phänomens (Tourism areas—Sociocultural constructions of a global phenomenon), Bielefeld: transcript 2010.
9. Lucius Burckhardt, *Warum ist Landschaft schön? Die Spaziergangswissenschaft* (Why is landscape beautiful? The science of strolling), Berlin: Martin Schmitz 2011.
10. Marc Augé, *Nicht-Orte (Non-Places)*, Munich: Beck 2010.
11. Traced in detail by Wolfgang Schievelbusch, *Geschichte der Eisenbahnreise* (History of the Railway Journey. On the Industrialization of Space and Time in the 19th Century), Frankfurt: Fischer 1993.
12. Stephan Oettermann, Berge weiten den Blick (Mountains broaden the view), in: Stephan Kunz et al., *Die Schwerkraft der Berge 1774–1997* (The Gravity of the Mountains 1774–1997), Basle: Stroemfeld/Roter Stern 1997, 49–55.
13. Erich Marx, 360 Grad: Vom Sattler-Panorama zum Location Placement (360 Degrees: From the Sattler Panorama to Location Placement), in: Kurt Luger & Franz Rest (eds.), *Alpenreisen* (Alpine Travels), Innsbruck: StudienVerlag 2017, 497–512.
14. On the art historical contexts, see Doris Hallama, Erhaben-bedrohlich-verbaut, Gebirgsbezwingung in der Kunstgeschichte (Sublime—threatening—built: mountain conquest in art history), in: Michael Kasper, Martin Korenjak, Robert Rollinger & Andreas Rudigier (eds.), *Alltag—Albtraum—Abenteuer* (Everyday—Nightmare—Adventure: mountain crossing and summit assaults in history), Vienna: Vandenhoeck&Ruprecht 2015, 205–222.
15. Marjorie H. Nicolson, *Mountain Gloom and Mountain Glory. The Development of the Aesthetics of the Infinite*, Ithaca: Cornell University Press 1959.
16. Matthias Stremlow, *Die Alpen aus der Untersicht* (The Alps from the Bottom View. Continuity and Change of Alpine Images since 1700), Berne: Paul Haupt 1998.

17. A grand cultural history of early alpinism from 1750–1850 is provided by Martin Scharfe, *Berg-Sucht* (Mountain Addiction), Vienna: Böhlau 2007.
18. The catalogue issued on the occasion of the exhibition in the Salzburg Residenzgalerie covers old and new images of the Alps. Erika Oehring (ed.), *Alpen—Sehnsuchtsort & Bühne* (Alps—Place of Longing & Stages), Salzburg: Residenzgalerie 2011.
19. Michael Jakob, Das Gebirge, das Heilige und das Erhabene (The Mountain, the Holy and the Sublime), in: Stephan Kunz et al., *Die Schwerkraft der Berge* (The Gravity of the Mountains 1774–1997), Basle: Stroemfeld/Roter Stern 1997, 75–81, 81. The author deconstructs in his fabulous essay the widely held view that alpinism started with Francesco Petrarca, because he had climbed this mountain "only for the sake of visual inspection". It is evident, that the letter in which Petrarca reports moved by his mountain tour, has been only written "at the threshold of the age of life to gravitas and looking back on his youth …", about 17 years after the ascent. This seems relevant to me, because even today inconsistencies in alpine literature attract public attention.
20. Michel Foucault, *Die Heterotopien* (The Heterotopias. The Utopian Body, Two Radio Lectures), Frankfurt: Suhrkamp 2005.
21. Klaus Kufeld, *Die Reise als Utopie* (The Journey as Utopia), Munich: W. Fink 2010.
22. Karlheinz Wöhler, *Touristifizierung von Räumen* (Touristification of spaces), Wiesbaden: VS-Verlag 2011.

4

A Brief History of Travel and Tourism

…It can probably be said that travels are by no means unimportant for any kind of life
Theodor Zwinger, *Methodus Apodemica, 1577.*

They are very enthusiastic about nature
And they promote interaction
They are very enthusiastic about nature
And only know the surroundings
From postcards.
Erich Kästner, *Vornehme Leute 1200 m hoch* (Distinguished People 1200 m high).

The history of mankind is also a history of movement, of mobility, of pushing out horizons. Initially, mainly crusaders, soldiers, messengers, and pilgrims were on the move, but mobility increased in the age of humanism. During the sixteenth century, metropolises of trade and science emerged, and business opportunities for traders expanded. Travel for religious, educational, or scientific reasons made

Rome, Naples, Paris, and Strasbourg focal points for many young nobles, scholars, and artists. Between these cities, the first fortified trail roads were established, precursors of today's road networks and the associated maps. However, it was not until the beginning of the nineteenth century that the systematic expansion of postal traffic networks through road construction using land-surveying techniques began in Europe.[1]

The usual means of transport during this epoch was the horse-drawn carriage, which was used on the postal routes. Where there were no tracks suitable for carriages, as on mountain passes, travellers were carried by locals in sedan chairs. Only the wandering vagrants, craftsmen on their way to a new job, travelling traders, or poets in search of adventure and self-affirmation went on foot. And then Johann Gottfried Seume, a writer from Saxony, chronicled his crossing of the Alps in *Walk to Syracuse*, published in 1802, heralded, like the wanderings of Jean-Jaques Rousseau, the beginnings of the path to today's nature-oriented tourism or alpine tourism.

The travel reports of this time are full of complaints about the inns and guesthouses, but also about the other travellers and their manners or behaviour. With the Europe-wide spread of the Ordinari-Fahrpost postal service at the beginning of the seventeenth century, the first travel compendia came onto the market, providing information about the road network, posting stations, and recommended accommodations. Among the many travellers who were on the move between Paris and Moscow, London, Stockholm, Rome, Hamburg, and Vienna in the eighteenth century, musicians probably have one of the oldest traditions. Long before travelling for educational or pleasure purposes became fashionable, they had been traversing Europe as minstrels entertaining princes, bishops, kings, citizens, and peasants. They did this at funfairs, imperial diets, at wedding and coronation celebrations, in pilgrimage sites, and at ecclesiastical councils. Mozart's letters to his family all full of complaints about the inconveniences of travel in this early period of globalized concert culture. To his "dearest sister," he wrote in August 1771 from Milan that he had endured "many heats" on the journey and "the dust has constantly impertinently dried us out". From Munich, he wrote to his "*mon très cher Père*" in Salzburg that the journey had been short but very arduous, the seats were hard as stone and he had not slept a minute at night, "this carriage jolts one's soul out".[2] The journeys of this time were certainly not pleasure trips; they were made for professional reasons, and the landscape to be traversed to the destination—as beautiful and untouched as it may have been—was of completely subordinate importance to these travellers.

Both passenger and postal transport were severely affected by war events. At the turn of the eighteenth/nineteenth century, for example, travel was hindered by the revolutionary events in France, the political upheavals that followed, and Napoleon's campaigns of conquest. Only after the 1815 Congress of Vienna did the postal administrations invest again in the further development of their services. Neglected paths and roads had to be improved, and paved country roads, specially built in Holland and France in the Baroque era to speed up traffic, were also established in other countries. With the developing express mail system, this resulted in a concept of highways, the importance of which went beyond comfort in individual transport and was linked to military interest. These roads significantly reduced the travel time of the stagecoach journey; moreover, because of the smoother surfaces, horse-drawn wagons could also transport heavier loads.

As early as 1754, a fast-moving *Journalière* service had been set up between Berlin and Potsdam. In England, faster passenger transport was carried out with the Mail Coaches, and in France with the Malle-Poste. However, speed came at the expense of comfort, as the idea was to transport people as quickly as letters. Reducing the time involved was possible only because the number of intermediate stops was cut drastically. In 1828, there was a daily travel connection between Berlin and Paris, and three times a week there was a fast post connection between Paris and Petersburg. The travel time on this route had been reduced from three to two weeks.[3]

Curiosity and Knowledge

The Austrian cultural sociologist Justin Stagl[4] refers to the epoch from late humanism to the scientific revolution—roughly from the mid-sixteenth to the mid-seventeenth century—as the age of "epistemological curiosity". It also brought forth the cultural technique of a refined art of travel, the *ars apodemica*. The focus here was systematic and methodical travel for scientific reasons, to collect the "treasures of wisdom and virtue scattered throughout the world". To distinguish proper travel (*peregrinari*) from aimless wandering (*vagari*), Stagl quotes Samuel Zwicker, who came from a scholarly family in Danzig, and his definition from the year 1577: "Travelling is, in fact, a change of place, undertaken by a suitable person out of a desire and a wish to cross, observe and get to know foreign places, in order to acquire there some good that may be useful either to the homeland and to friends, or to oneself."[5]

It was during this period that the refinement of the three cultural techniques of *travel, surveys, and systematic collection*, which would later become the methodological canon of social research, began.

In the late eighteenth century, the myth of the ideal total traveller with his scientific claim faded, but travel more focused on the individual's sensations, and the corresponding travel literature, continued.

While the Dutch university town of Leiden was an obvious travel station for the scholarly journey, the imperial city of Vienna was such for the noble journey in the age of absolutism. The Grand Tour of the young heirs of established and landowning noble families were undertaken for the purpose of grooming them into the appropriate lifestyle and manners of the European aristocracy. Under the supervision of a tutor, these young cavaliers moved between the most important cities, castles, and universities by carriage. France was visited for its gallant manners, fashionable elegance, and sociable interaction; in Italy the focus was on the architectural testimonies of antiquity, the courtly art of opera and comparative political science. For the Catholic nobility, Rome was the highlight of the journey, The Netherlands and England were visited because of their higher-ranking economic-technical development, and Vienna was an indispensable destination because the imperial city could serve as both a test of what had been learned so far and a springboard for future use in royal services for the young nobles. Court life and cosmopolitan experience, two main features of noble travel experience, were closely linked.

And then, in the second half of the eighteenth century, walking—a slow, almost tactile movement through landscape and society—developed as an early form of what was to become today's hiking tourism. It served as a means of collecting physical and sensory experience through personal observation, following the horizon of ideas of the Enlightenment. Noted and written down, the seen was organized and brought into a system of experience. This marks the beginning of the great era of travel literature, which foregrounds the self-discovery of one's own self in the foreign as an important school of bourgeois character education. However, walking also had an anti-feudal aspect and embodied an attitude that differed symbolically from the debonair Grand Tour of the young nobles, particularly in the German principalities and kingdoms, an element of the bourgeois spirit of opposition against the noble privileges of late absolutism, expressing a social partisanship for the people.

Forms and Phases of Tourism

If we try to systematize tourism, to bring it into a chronological order of development, we may say that the first epoch of travel ends with the noble and later bourgeois-influenced individual journeys. Although the Grand Tour was undertaken primarily for other motives, it had clear elements of early educational and cultural tourism as well. The places visited were the ancient art sites, churches, monasteries, treasure and art chambers, and similar attractions throughout Europe. The bourgeois educational journey emerging with the Enlightenment served to broaden the horizon in every sense, as regards Italy and later Greece. Travel became a new form of world experience, inspired by the reading of great writers such as Goethe's travels to Italy in the late eighteenth century, which triggered longings for Italy north of the Alps.[6] The usual means of travel were the carriage and the sailing ship. The roads were poor, and the travel speed was low.

Already in the fifteenth century came the first world travels of explorers, soldiers, and scientists, financed by the monarchs and merchants. Europeans ventured into the totally unknown continents, took possession of them, and subjugated the local populations. Since the sixteenth century, Jesuits, like the traders, had travelled along the

Silk Road to the court of the Chinese emperor. The Jesuit Father Grueber from Linz, today Upper Austria, reported to his client, the Pope, about living conditions at the court of the Chinese emperor and the weapon arsenals in the Middle Kingdom.[7] The emergence and expansion of colonial empires overseas and the tales of exotic islands and wondrous regions aroused interest among broad social circles. Adventurers crossed the deserts and mountains on horseback or with camels, searching for the sources of the Nile, and finding a flourishing slave trade across the dark continent. Towards the end of the nineteenth century, the bourgeoisie's desire to travel blossomed due to world exhibitions arranged in European capitals. The travel literature of the time was shaped by the superiority of European civilization, as verfied in the academic writings of Edward Said on *Orientalism* and *Culture & Imperialism*.

In the interwar years, daredevils undertook the first long-distance travels with motorcycles and cars, and expeditions explored the high mountains of South America and Asia. These travel reports and photos captivated audiences, just as the alpinists' struggle for the summits—and with death. The intense longing for world experience, for mobility and encounters with other peoples, as well as for entertainment and distraction, ultimately led to a global travel market after the years of wartime deprivation, becoming one of the world's largest economic sectors by the second half of the twentieth century.

However, even today, we need to realize that travel as a self-evident part of culture or a way of life remains a privilege of the wealthier in the Western world. The most beautiful nooks and crannies of the poor countries of the global South have indeed become exclusive destinations for long-distance tourism in the course of market diversification—but only small segments of their population can afford holiday travel to this day.

Costal and Bathing Tourism

Around the middle of the eighteenth century, spa tourism or the turn towards the sea began, first in Great Britain, and towards the end of the century also on the European continent. Bathing tourism in Scarborough started as early as the seventeenth century. In the second half of the eighteenth century, the foundations for the fashionable bathing of upper-class English society were laid in the small fishing village of Brighton. Spa promenades and seawater bathhouses were built, and Brighton was upgraded by the plans of the English royal house to build a summer residence here. On the mainland, the first sea bathing establishments were created on the French Atlantic coast and the English Channel, a little later on the Baltic Sea.

The sea, untamed nature, was deemed to have powers to restore harmony between body and soul, to counteract the loss of life energy and the widespread melancholy and restlessness. The sea lust stimulated by such expectant curiosity—as noted by Alain Corbin[8] in his comprehensive study of the Occident and the 'discovery' of the coast—stimulated elite society to travel to the sea, seeking to remedy the negative effects of urban civilization, which were already perceived as harmful at that time. It was believed that a stay at the sea would heal the mind overstrained by too much thinking; the seaside toughens spoilt urban dwellers. The oceans embody the forces of untamed nature that tolerates no lie. The sea became a refuge—giving hope precisely because it instils fear and a sense of wonder.

Also today, holidaymakers expect beneficial effects on their health from stays at the sea, where sky and water form the horizon, together delivering a package that has become the epitome of a successful holiday.[9] Such warm-water destinations have come to account for the vast majority of all holiday trips, as reflected in the almost seamless development of monocultural tourism landscapes.[10]

Around the same time as the discovery of the sea coasts, trips to inland springs became fashionable, and the first spa towns emerged where healing springs bubbled and promised relief for a range of ailments. These spa stays were initially a privilege of the nobility and the wealthy bourgeoisie, but by the end of the nineteenth century, also the middle classes could afford them. Many spa towns have been able to preserve their characteristic appearance to this day—like Bath in southern England, a World Heritage site since 1987, Karlovy Vary in today's Czech Republic, as well as Bad Gastein and Bad Ischl in Austria. All these places benefited from the stays of crowned heads as well as famous musicians and writers, who imparted an intellectual and artistic flair. Fountain and promenade halls, casinos, luxury hotels as well as spa parks emerged: the classic spa town.[11] In 2021, eleven significant European spa towns were added to the UNESCO list as transnational and serial World Heritage *Great Spa Towns of Europe*.[12]

Similar to this type of health tourism are the *Sommerfrische* (summer retreat) that emerged in the mid-nineteenth century: wealthy bourgeois families would stay in rural villas or guesthouses to escape the oppressive summer heat of the cities. During the time of the Austro-Hungarian Monarchy—especially in the second half of the nineteenth century—the aristocracy and imperial officials spent the summer months near His Imperial Majesty, who stayed in spa towns in the Salzkammergut or the Vienna Alps around the Semmering. Also today, the increasingly frequent hot summers due to climate change, even in temperate latitudes, have continued to boost summer retreat tourism and the summer season in the Alps.[13]

Phases of Tourism Development

The *first phase* of modern tourism, 1835–1880, began with the age of the railway, which allowed large numbers of people to be transported: the beginnings of a tourist infrastructure were created. Spa towns and seaside resorts expanded, hotels and guesthouses were built, travel agencies were established to handle travel business and bookings. Tourism as a phenomenon of modernity stems from industrial transformation, arising in England, and later expanding to the entire Western European industrial core. However, this applied originally not to the working classes but to the urban bourgeoisie, who could now afford holiday travel—once the purview of the nobility, who in turn came to seek new destinations.

Already at the beginning of the nineteenth century, members of the nobility began venturing into the high mountains: the first Alpine Club was founded in London as a Gentlemen's Club in 1857. The Austrian Alpine Club was established in Vienna in 1862, and the German Alpine Club in Munich in 1869. First ascents in the Western and Eastern Alps followed, and the members of the Alpine Clubs sprinkled the mountain world with alpine huts and paths, aimed primarily at promoting mountaineering and youth hiking. A tourism infrastructure was established in many Alpine valleys, and with the expansion of the railway network, the cities moved closer to the mountains in terms of time.[14]

In 1845, Thomas Cook founded his first travel agency in England, which he developed into a worldwide organization offering package tours. Benefiting from the many overseas possessions of the British colonial empire, he became the world's most famous tour operator.[15] (However, the world's oldest travel company declared insolvency in 2019). In 1876, the Compagnie Internationale des Wagons-Lits (CIWL) was founded in France, which with its luxury trains like the *Orient Express* captured much of the train travel market and, alongside Cook and American Express, becoming a major travel agency.

Germany followed suit: the travel agency *Stangen* adopted Cook's package system, whereby tourists were ferried by ship, train or stagecoach and accommodated in booked inns along the fixed travel route—for a fee payable in advance.

In the first half of the nineteenth century, the first travel guidebooks appeared on the market. In 1828, the German Karl Baedeker published the travel description *Rhine Journey from Mainz to Cologne*. His travel guides offered detailed and reliable information in a simple and factual language. *Murray's Handbooks for Travellers* appeared from 1836, offering precise travel plans for destinations in Europe as well as parts of Asia and North Africa. This reflected the emerging tourism industry and the British trade and industry organization in general, offering both practical guidance for travellers and security.[16]

With the establishment and expansion of summer retreats and forms of travel appealing to the middle social classes, we can speak of a *second phase* of tourism history that extends until the First World War. At the end of the nineteenth century, with the growing industrial proletariat in many countries, a labour movement emerged that increasingly became an important political actor. It fought not only for the rights of workers such as higher wages and better working condition, but also for shorter weekly working hours; Sundays and holidays became non-working days, thus offering broad opportunities for Sunday outings. The workers' associations also saw themselves as an educational movement and workers' leisure organizations. In 1895, the proletarian tourist association Die *Naturfreunde-Friends of Nature* was founded in Vienna. The destinations were usually not very far from the places of residence and could be reached by train or bicycle. What was to become the UK-based *Holiday Friendship* (HF) was founded by a British pastor in 1891.[17]

The first mass tourism phenomena—as a *third phase* of development—date back to the 1930s. During the First World War, all tourism had come to a standstill. In 1925, the Opera Nazionale Dopolavoro (National Organization after Work) was founded in Italy, a true production facility workshop of fascism, through which leisure time was modelled and controlled under uniform national aspects. It was to be consumed only exclusively collectively. Hikes, marches, shooting exercises, folkloric social trips and mass events were organized to generate a fascist national community. In the German Reich, with the help of the National Socialist *Kraft durch Freude* (KdF, Community Strength through Joy), a politically organized people's and social tourism was created following the Italian model. The political control of people aimed to cover cultural and social contrasts through common leisure use and to develop a sense of belonging—one people, one empire. The NS community KdF thus became the largest travel organizer at the time, selling 40 million trips between 1934 and 1939. Tourism historian Hasso Spode calls this the first German travel boom.[18]

In the decades after the Second World War, after the difficult years of reconstruction, a leisure society gradually emerged from the working society.[19] Vacation became a part of lifestyle at least in Western societies, annual leisure time increased enormously and the number of working days gradually decreased. Reasons for this were the advancing mechanization and technologization of work processes as well as social policy reforms.

The complete industrialization of tourist travel and travel for every man & his wife marks the *fourth phase* of modern tourism, which began in the Western world around the 1970s and ended with the Corona pandemic in spring 2020. Since the turn of the millennium, particularly the emerging countries of Asia have developed a strongly expanding tourism that also significantly reflected in arrivals in Europe. The number of cross-border trips in 2019 was about ten times as high as 50 years ago. In 2018, Europe recorded a total of 740 million tourist arrivals, more than twice as many as the Asia–Pacific region with 360 million and more than three times as many as America with 220 million. The top 10 destinations account for 40% of worldwide tourist arrivals — France 89 million, Spain 83, USA 80, China 63, Italy 62, Turkey 46, Mexico 41, Germany 39, Thailand 38, UK 36 and the small landlocked Austria is just behind with almost 30 million. The countries with the highest international tourism expenditures were China, the USA and Germany by a large margin. Europe accounts for approximately 40% of all international tourism expenditures, followed by the Asia–Pacific region and the American countries. 51% of all international tourism trips are to Europe, a quarter to Asia–Pacific and 15% to America. Africa and the Middle East together account for about 10%. 56% of all trips are holiday or pleasure trips, 27% are for health or religious reasons, 13% are for professional reasons. The share of air travel has increased from 46% in 2000 to 58% in 2018, roughly the same amount as land tourism has decreased. In 2019, approximately 47 million flights were recorded in global aviation, with air travel revenue amounting to 612 billion US dollars. Triggered by the travel restrictions due to the Corona pandemic, the global market shrank by around 80%.[20]

Tourist travel in Europe is predominantly by car. Coaches and trains are of secondary importance, airplanes have become a matter of course not only in long-distance tourism. Cruises have particularly gained in popularity. Worldwide, 28.5 million passengers took a cruise in 2018, a doubling compared to 2005. The desire to travel among the urban population is much more pronounced and to this day the upper social classes travel more frequently and for longer periods.[21]

The Touristification of the Globe

The further development of transportation and the widespread development, allowing for the touristification of almost all attractive regions for tourism, have significantly contributed to this global development. While the railway had initiated the first major expansion and was still the dominant mode of transport after the Second World War, today in postmodern tourism it is clearly the car, as it extends the radius of action and allows for even greater individual mobility. Its unbridled development is the central characteristic of the highly individual tourism of the last decades, although the tourism destinations offer a wide variety of standardized services. The rapidly growing motorization of the population and the expansion of the road network contributed to this shift in tourism, as even more remote destinations became easily accessible, whereas in the past only areas within the catchment area of the railway were affected by tourism.

Today, about three out of four Germans travel by car to their holiday destinations, and in other European countries it is not much different, especially since for family vacations the car is the most cost-effective means of transport. Camping only became a tourism trend through the car, and from the tent to the caravan to *glamping*, a deluxe camping, and to rolling living rooms, a tourism sector directly oriented towards the car was created. The bus as a means of travel has lost importance compared to the car, but it continuously successfully serves the group travel segment. In times of climate change, it has gained in image because it emits significantly less greenhouse gases compared to individual car traffic and is therefore considered more environmentally friendly. The largest negative ecological footprint is left by air travel followed by cruise ships. The airplane has become the most important means of transport for package tourism, which relies on mass production, standardization, division of labour, and high unit numbers. Charter trips from the agglomerations to the holiday destinations are now part of everyday tourist life. With the introduction of so-called low-cost carriers or due to the drop in prices for air travel, short and long-haul flights have become affordable for a broad section of society. The ongoing boom in cultural and city tourism is partly due to this.[22]

Tourist car traffic not only has negative effects due to environmental pollution, but also due to traffic density. The mass migrations in slow-moving traffic over the congested highways to the holiday regions on the Mediterranean or in the Alps create similar crowding effects in the destinations. In city tourism or in some ski resorts, people talk about *overtourism*, a phenomenon where the capacity of a destination is overstretched and the quality of experience for travellers as well as the quality of life for residents is impaired.

Since the turn of the millennium at the latest, we can speak of a global tourism industry that uses all means of transport and has also developed a pleasure periphery of even the most remote regions. The multifarious sectors of tourism cater to all needs that arise from the deficits of everyday life—and which essentially also represent the motives for travel, such as the desire for self-realization, relaxation, pleasure, prestige, and the search for temporary happiness.

Today's modern, hurried people have the permanent feeling of missing out on something—and in the process, have lost sight of the beauty of being 'underway', which calls for a pause.[23] *Slow Tourism*—slow forms of vacation and travel such as nature-based groups of car-free holiday resorts and mountaineering villages have emerged as niche markets because they meet the need for a slower pace. Karlheinz Wöhler sees in the detachment from everyday life therefore not an escape, but rather the maintenance of the self in everyday life, because travelling interrupts the profane everyday life and gives new meaning to life. Consumer, leisure and tourism worlds are guiding spaces of postmodernity, which need an everyday space to be perceived as another world. Removing oneself from familiar spaces allows for being different, for being able to go beyond oneself. Tourism thus becomes something out of the ordinary and in spaces of desire, people learn to recognize the incompleteness of their being. In tourism spaces, therefore, places are suspected where the meaning of life unfolds and experience deficits can be remedied—they are happiness-making distant spaces, *heterotopias*.

However, temporary spatial stay is also a consumer good that is economically used as a mythical form or as a beautified space. In mass tourism, the landscape has become a consumer item and rooms with a view are sold more expensively than those in the backyard—this applies to the mountains as well as to the sea or a historic old town. Such panoramas of beauty form the basic capital for a tourism landscape, where the picturesque and the typical are traded as landmarks of a region or even as national symbol items. The Bay of Capri, the steep flanks of the Matterhorn, the street canyons of Manhattan, the view from the Trocadero over the Champ de Mars with the Eiffel Tower—they are icons and recognition images, with which the longing for experience through personal observation and intensity of experience is triggered or increased.

Tourism in the digital age is exploration with views that no longer need to be defined by one's own eyes. Art has practiced the role of the scout, for example in landscape painting, creating images that are enriched in the cross-media tourism marketing of experience space management into stereotypical image brands and further experience promises. The experience of the landscape by means of railway, cable car, or car thus triggers a tension between nature and technology, as this transforms a wild and inaccessible mountain landscape into a prototypical tourism landscape, into a transformed landscape of pleasure and experience. The new landscape feeling takes into account the desire to technically dominate nature.[24]

Examples of such transformation processes include the railway line over the Semmering. It has been a World Heritage Site since 1998 due to its outstanding technical achievement as the first high-mountain railway and because it made areas of great natural beauty more accessible, opened up for residential construction and recreation, leading to the creation of a new form of landscape. The same applies to the Rhaetian Railway in the Swiss Alps, a World Heritage Site since 2008 due to its impressive route and construction technique. The same is true for daring routes and alpine roads such as over the Dolomite passes, the Grossglockner High Alpine Road from Salzburg to Carinthia through the Hohe Tauern National Park or the Norwegian National Scenic Routes, but also for Ocean Drives like the Corniche on the Côte d'Azur or the Amalfitano, the winding coastal road along the Gulf of Salerno south of Naples.

The accompanying transformation applies to many other landscapes on all continents. After the invention of the car came the panoramic roads, the parkways in the

USA for example, because the city dwellers wanted to explore nature and the periphery of the cities with their motorized vehicles. The masterpieces of engineering of the nineteenth and early twentieth centuries thus not only document the impressive technical progress, but can also be seen as symbols of the taming of nature and its preparation for a cinematic viewing pleasure for visitors. They now form a nature theatre that can be perceived through enjoyable consumption. The use of tourist infrastructure creates a tourist space that is experienced through this mode of relationship. Its traversal emotionally reduces the experiential space and shifts concepts such as proximity and distance.

Endnotes

1. A good overview is provided by *Reisekultur* (Travel Culture, From Pilgrimage to Modern Tourism), edited by Hermann Bausinger, Klaus Beyrer and Gottfried Korff, Munich: C. H. Beck 1991.
2. Stefan Kunze (ed.), *Wolfgang Amadeus Mozart. Briefe* (Letters), Stuttgart: Reclam 2005, 47 and 163.
3. Hermann Glaser & Thomas Werner, *Die Post in ihrer Zeit* (The Post in its Time, A Cultural History of Human Communication), Heidelberg: Decker 1990.
4. Justin Stagl, *Eine Geschichte der Neugier* (A History of Curiosity, The Art of Travel 1550–1800), Böhlau: Vienna 2002, as well as Bausinger, Beyrer and Korff, Munich 1991.
5. Stagl, 2002, 5.
6. Here and in the following, detailed by Rüdiger Hachtmann, *Tourismus und Tourismusgeschichte* (Tourism and Tourism History), in: Docupedia-Zeitgeschichte (https://doi.org/https://doi.org/10.14765/zzf.dok.2.312.v1, accessed 20.04.2021).

4 A Brief History of Travel and Tourism

7. Johannes Grueber, *Als Kundschafter des Papstes nach China* (As a Scout of the Pope to China 1656–1664), Stuttgart: Erdmann 1985.
8. Alain Corbin, *Meereslust, Das Abendland und die Entdeckung der Küste* (Sea Lust, The Occident and the Discovery of the Coast), Frankfurt: Fischer 1994.
9. Marine biologist Wallace J. Nichols sees the close relationship between water and the human brain as a reason for this well-being. *Blue Mind, How Water Makes Us Happy*, New York: Brown 2014.
10. For the development of tourism on the Mediterranean coasts or on Mallorca see Andreas Kagermeier, *Tourismusgeographie* (Tourism Geography), Konstanz: UTB 2016.
11. Gabriele Knoll, *Kulturgeschichte des Reisens* (Cultural History of Travel, From Pilgrimage to Beach Vacation) Darmstadt: Primus 2006.
12. The transnational site of The Great Spa Towns of Europe comprises 11 towns, located in seven European countries: Baden bei Wien (Austria); Spa (Belgium); Františkovy Lázně (Františkovy Lázně, Czechia); Karlovy Vary (Czechia); Mariánské Lázně (Czechia); Vichy (France); Bad Ems (Germany); Baden-Baden (Germany); Bad Kissingen (Germany); Montecatini Terme (Italy); and City of Bath (United Kingdom). These sites display the significant interchange of human values and developments in medicine, science and balneology. See also: https://whc.unesco.org/en/list/1613, 05.08.2021.
13. Hanns Haas, Die Sommerfrische (The Summer Retreat—A Lost Tourist Cultural Form), in: Hanns Haas, Robert Hoffmann & Kurt Luger (eds.), *Weltbühne und Naturkulisse* (World Stage and Natural Scenery—Two Centuries of Salzburg Tourism). Salzburg: Pustet 1994, 67–75; Wolfgang Kos, *Der Semmering—Eine exzentrische Landschaft* (The Semmering—An Eccentric Landscape). Salzburg: Residenz 2021.
14. Rainer Armstädter, *Der Alpinismus—Kultur, Organisation, Politik* (Alpinism—Culture, Organisation,

Politics)., Vienna: WUV Universitätsverlag 1995; Anneliese Gidl, *Alpenverein, Die Städter entdecken die Alpen* (Alpine Club: City dwellers discover the alps), Vienna: Böhlau 2007; Katharina Scharf, *Alpen zwischen Erschließung und Naturschutz, Tourismus in Salzburg und Savoyen 1860–1914* (Alps between development and nature protection), Innsbruck: StudienVerlag 2021.

15. Jörn Mundt, *Thomas Cook—Pioneer of Tourism,* Konstanz: UVK 2014.
16. Hasso Spode, Der moderne Tourismus—Grundlinien seiner Entstehung und Entwicklung vom 18.–20. Jahrhundert (Modern tourism), in: *Moderner Tourismus—Tendenzen und Aussichten, Materialien zur Fremdenverkehrsgeographie.* Trier: Geographische Gesellschaft 1992, 39–76.
17. Manfred Pilz, *„Berg frei". 100 Jahre Naturfreunde* (100 years Friends of Nature), Vienna: Verlag für Gesellschaftskritik 1994; https://www.hfholidays.co.uk/about-us/about-hf-holidays/timeline, accessed 8.8.2021.
18. Hasso Spode, Die NS-Gemeinschaft 'Kraft durch Freude' (The NS Community 'Strength through Joy'—a people on the move?), in: *Zur Sonne, zur Freiheit* (To the Sun, to Freedom! Contributions to the History of Tourism), Reports and Materials No. 11 of the Studienkreis Tourismus und Entwicklung and the Institute for Tourism of the FU Berlin, Berlin 1991, 79–93.
19. Hans-Werner Prahl, *Soziologie der Freizeit* (Sociology of Leisure), Paderborn: Springer 2002. In Germany, the tourist-oriented empirical leisure research was shaped in the 1990s by Horst Opaschowski, who conducted a series of studies with the demoscopic B.A.T Institute and published extensively. For his overview article on 'leisure psychology' see Heinz Hahn & H. Jürgen Kagelmann (eds.), *Tourism Psychology and Tourism Sociology: A Handbook for Tourism Science*), Munich: Quintessenz 1993, 79–84.

4 A Brief History of Travel and Tourism

20. According to the UN-WTO and country-specific differences. See https://www.unwto.org/international-tourism-and-covid-19, accessed 31.07.2021; and https://www.e-unwto.org/doi/book/https://doi.org/10.18111/9789284421152, accessed 31.07.2021.
21. https://de.statista.com/themen/702/tourismus-weltweit/, accessed 31.07.2021.
22. A comprehensive overview is provided by Sven Groß, *Handbuch Tourismus und Verkehr* (Handbook Tourism and Transport—Transport Companies, Strategies and Concepts), Konstanz: UTB 2017.
23. Vom Verlassen der Paradiese, (From Leaving the Paradises. The insurmountable romantic's philosophical perspective on travel, including tourism), in: Roman Egger & Kurt Luger (eds.), *Tourismus und mobile Freizeit* (Tourism and mobile leisure—lifestyles, trends, challenges). Norderstedt: BoD 2015, 11–26.
24. Bernhard Tschofen, *Berg-Kultur-Moderne* (Mountain-Culture-Modernity, Folklore from the Alps), Vienna: Sonderzahl 1999; Thomas Zeller, *Straße-Bahn-Panorama* (Road-Rail-Panorama, Transport routes and landscape change in Germany from 1930 to 1990), Frankfurt: Campus 2002; Thomas Zeller, *Consuming Landscapes. What We See When We Drive and Why It Matters*, Baltimore: Johns Hopkins 2022.

5

Utopias and Dystopias—Dream and Nightmare

We all know this: Some landscapes makes the heart beat faster:
A sunset that moves the spirit, a view that leaves us Speechless.
For one It is the mountains, for the another the sea, for yet another the desert.
The senses suddenly react more keenly, more sensitively.
Colours glow differently, distances change, boundaries Shift.
Sometimes it is the unexpected, the new, that comes into view.
Or it is the return, the reunion, that captivates.
Iso Camartin, *Jeder brauch seinen Süden* (Everyone needs their South).

Last year, I was on a round the world tour with my wife.
Listen, I'm not going there anymore.
Gerhard Polt, Bavarian cabaret artist.

Setting off, seeking new things, discovering the world and oneself in the process—whatever the motives for travelling to countries near or far may be, hardly any region embodies this longing more than the mountain world of the Hindu Kush-Himalaya, twice as high and twice as long as the Alpine massif. Travellers, wanderers and climbing tourists seek and find the places of their desired happiness there. Their rapt tourist gaze makes them close their eyes to what is ugly, blocking out the threatening reality. Tourists do not want to hear about the horrors of the merciless monsoon rains or the rapid demise of the glaciers. They only want to see the beautiful, which is still there in abundance: the great silence and sublimity of the mountain world, tiny villages that huddle like flocks of sheep on the green hills, handcrafted and cultivated terrace fields that blend harmoniously into the wilderness, flora and fauna as colourful as a bird of paradise, and fascinating, ethnically and culturally mixed populations. The shy grace of a smiling peasant girl, the warmth of the farmers—even their poverty seems charming, and the spirituality of this landscape hovers above the white peaks like a gentle cloud.

The Garden of Eden—A Lost Paradise

Life in the world's highest mountain range can be very hard, devoid of romance. And if, as Marcel Proust believes, the lost paradise is the only paradise, then the Himalayas are indeed an incarnation of a Garden of Eden—but one whose destruction is already well underway. This concerns not only the ruthless exploitation which has long since engulfed the entire planet and is throwing the ecological balance into disarray everywhere. In view of the other threats to which the world's highest mountain range is exposed, it is no exaggeration to speak of *places of hell*. Wars, civil wars, terror, poverty, hunger, discrimination and violent racism, catastrophic government performance, corruption, destruction of biodiversity and ancient cultural practices—all this can be found

5 Utopias and Dystopias—Dream and Nightmare

in these countries. And yet, the Hindu Kush-Himalaya region persists in the fantasies of Western tourists as a destination of longing: childhood dreams of a paradise on earth are spun further, ultimately condensed into travel plans and travel decisions now within reach, thanks to the low airfares that make long-distance travel affordable for many.

Firmly established as Shangri-la in the glossy magazines of the travel industry, this region usually makes the international headlines only when an event stands out glaringly—whether because of its curiosity, its absurdity or its incomprehensibility and brutality. One such occasion occured when in the spring of 2013 an Islamist terror commando on the fairytale meadow of North Pakistan's Nanga Parbat, the ninth highest mountain in the world, pulled a group of mountaineers out of their tents and murdered them in cold blood. By contrast, the deaths of hundreds of mountain farmers almost every year, caused by the increasingly unpredictable monsoon, remains unnoticed, or merit only a brief mention in the back pages.

"If there is a paradise on earth, it is here, it is here," raved the great Mughal emperor Jahangir more than 350 years ago. Peaceful and enchanting, the Vale of Kashmir lies in all its splendor at the foot of the snow-covered mountain ranges of the Himalayas and Karakorum. The Jhelum River winds like a silken thread through the lush green paddyfields; in shimmering Dal Lake the houseboats are moored and children greet strangers with lotus flowers, which bloom by the thousands in the floating gardens around the city of Srinagar—named after *Lakshmi*, the divine goddess of prosperity, wealth and abundance.

However, this idyll has one tragic flaw: Pakistan and India remain in dispute over Kashmir, the beautiful bride of Asia, divided since 1947. The hostile neighbours have already fought several wars, and a lasting peace is not in sight, as both sides continue to fan the flames. The calls of the Muslim majority for an *Azad Kashmir*, a united and free Kashmir, are getting louder, but India does not want to listen. In addition to this long-standing conflict, Pakistan is one of the world's most violence-ridden countries, serving as a retreat and staging ground for fundamentalist Islamists and terrorist groups. And yet, the rural population still practice a tradition of generous hospitality hardly found elsewhere.

Poverty, destruction of nature, suppression of traditional ways of life and cultures characterize the situation in many parts of the Hindu Kush-Himalaya. In Afghanistan, Kashmir and until recently Nepal, people live in war-zones riddled with violent ethnic, political and religious conflicts. In the Indian part of Kashmir, the front line extends up to 6000 m. On both sides of the Siachen glacier, heavily armed soldiers face each other. China has declared the vast Tibetan highlands a military buffer zone and uses it as a nuclear waste and weapons depot, and pursues a strict policy of 'Sinicization': the political, economic and cultural integration of Tibet into the Chinese social system. Neighbouring Nepal suffered a brutal civil war that lasted ten years, until the Maoist rebels ultimately overthrew the king with the help of parliament—enjoying wide backing, because the situation of the farmers was so desperate and the corruption of the rulers so ruthless that they wanted to get rid of this system of oppression at all costs. In the young republic, however, despite a new federal constitution, much still needs to change before we can speak of a real socio-political change. Corruption has not really decreased: now the big limousines are driven by other *fat cats*, as the Nepalese call their ruling upper class.[1]

Despite the many violent conflicts on the roof of the world, despite the natural disasters and the massive threat to the ecological balance, international Himalaya tourism has grown constantly: the magic emanating from this region

5 Utopias and Dystopias—Dream and Nightmare

seems unbroken. Domestic tourism and pilgrim numbers are also increasing. Instead of going to Kashmir, residents of the Indian metropolises now travel to other northern Indian states during the hot summer months—to Himachal Pradesh, Sikkim or Darjeeling. And tourism in Tibet has been booming for years. Lhasa is experiencing a rush of pleasure-seeking mainland Chinese—and the Old Town is witnessing the decline of its sacred culture. The centre with its fabled holy places has become a Disneyland folklore district with all the trademarks of mass tourism. The World Heritage show business is flourishing: the Potala, once the residence of H.H. the Dalai Lama, is being trampled down, while traditional old town houses are falling into disrepair or being mercilessly modernized. The 2000-km railway line from Golmud across the high plateau to Lhasa, opened in 2006, shovels more and more tourists to Tibet; the expansion of the connection to Shigatze took place at a rapid pace as well. Already planned is the continuation to the west to Mt Kailash, the holiest of the holy mountains and pilgrimage site of many Himalayan peoples, and beyond to Ürümqi in the Uighur Autonomous Region Xinjiang (East Turkestan). Until recently still a hinterland province, Tibet is now being developed infrastructurally at a rapid pace, preparing for mass tourism.

In Nepal too, tourism had been growing steadily, with the conquest of the highest peaks and hiking tourism in the shadow of the peaks higher than 7000 m., indeed, even 8000 m. Then the after-effects of 9/11, the Nepali civil war and a destructive series of earthquakes in 2015 led to a massive decline. The number of trekkers and mountaineers was halved after the civil war—but ten years after this violent wildfire, almost a million tourists were recorded. More recently, however, tourism came to a near standstill due to the Covid pandemic, as would-be visitors could no longer enter the country, or be permitted to return home from there.

However, most visitors do not get beyond the capital Kathmandu. Pilgrims visit the local holy sites of Hindus and Buddhists; shopping- and gambling-addicted Indians and Chinese swarm to the shopping malls, casinos and similar establishments. In Kathmandu's tourist district Thamel, a Little Chinatown has emerged, with hotels, restaurants and entertainment. For Chinese private companies, these are profitable investments. Geopolitically motivated, the Chinese state also invests in Nepal's infrastructure, in roads and hydroelectric power plants: this also benefits tourism. Western visitors to Nepal are mainly interested in culture, in the sacred World Heritage architecture in the Kathmandu Valley and in the unique mountain landscapes. This mixture of paradisiacal subtropical landscape, ethnic and religious diversity, affordable service and captivating friendliness of the locals, complemented by an overwhelming offer of exotic and cheap souvenirs, has made the country the centre of Himalaya tourism.[2]

Paradise is Always Elsewhere

Tourism thrives on dreams of paradise—and not just in the Himalayas. The Salzburger Land has for many years attracted visitors with its enchanting alpine landscape, cheekily calling itself the *Little Paradise*, while a premium hotel chain is called *Shangri la*. Otherwise, the world map of long-distance tourism features paradises for divers; under exotic palm trees, Tonga is the sleepy paradise of the South Seas, and island states like Nauru and Kiribati are among the 10 last relatively tourist-free paradises.

In the Himalayas, the ancient kingdom of Bhutan is billed as the last paradise. With its white peaks, dense

5 Utopias and Dystopias—Dream and Nightmare

forests, and Mahayana Buddhist culture, the small state is the tailor-made answer to Western longings. This paradise admits only a small circle of chosen visitors, keeping the numbers artificially low and the prices high. For years, Bhutan has been the *enfant chérie de la terre*. Generously supported by European development aid, the modernization of the country is proceeding with caution under the slogan *Gross National Happiness*. In reality, geopolitically motivated Bhutan is no longer a 'developing' country. Governed by a young generation of technocrats loyal to the king and educated in Western universities, Bhutan has established itself as an exclusive tourism destination. Its youthful fairytale king even allows a parliament—which, however, consists solely of politicians loyal to the king. This is storytelling and public diplomacy supreme: a global marketing and PR strategy diverts attention from the fact that Bhutan has probably produced the world's highest number of ethnic refugees per capita. Gross National Happiness is only for the majority population of the Drukpas, not for the Nepalese-origin Lhotsampas. Over 100,000 of them have been expelled from Bhutan since the 1980s, accused of being illegal foreigners. And yet, representatives of supranational organizations rave about this model country, which protects its natural heritage and preserves cultural traditions.

Bhutan's politics could be understood as a joint venture with the West: It chants the mantra of anti-materialism and incorporates its nature into a tourism concept that only wealthy happiness-seekers from the West can afford, from where also the development funds and ecological know-how come.[3]

Utopian Imagination Paradise

Paradises are imaginative concepts that exist in almost all religions and cultures. As paradise—the word comes from Old Persian—the Islamic peoples celebrate the most beautiful gardens in their poetry. In the poetry of the Sufis, bliss and harmony thrive, and there is boundless joy in this flawless nature.

In literature, painting and handicrafts, trees, mountains and water are constitutive components of natural beauty and such paradisiacal landscapes. In the Christian paradise myth, spatial phantasmagorias are linked with heavenly spheres, with purity and divinity. Many of these elements are found as building blocks of a tourism marketing deliberately focused on awakening such longings. Vacationers are lured with promises of sunshine, blue skies, tropical beaches, crystal-clear waters, flawless bodies, exoticism, untouched landscapes, sensual abundance—a timeless world in which peace, harmony, and prosperity exist, a truly blissful state.

The first tourist paradise images to be spread via modern mass media featured Hawaii, the island world of the South Pacific, the South Seas—imaginary geographies that have been slumbering in collective illusions since the records of the early circumnavigators. Travelling means following images and experiencing on site what we could only dream of before. Travelling, we consume the goals of our desires—and are disappointed when reality—the real Tibet, Tyrol, Taormina, Tahiti, etc.—fails to keep up with our imaginings. With the increasing integration of tourism into the culture industry, the fashion industry responded to ethno-vestimentary needs and the audiovisual industry increasingly shaped the imaginary worlds of those seeking relaxation and distraction with a nomadic temperament in Cinemascope and on the home screen. Today's large-scale flat screens and smartphones increase the magic with brilliant colours showing a wealth of always-accessible dream-places.

The magnificent mountain world of the Himalayas also exists in the imaginations of many as a destination of longing. Around it, myriad legends and myths are woven, those of the conquest of the highest peaks as well as those of the wisdom and spirituality of Buddhist and Hindu demons and gods. Images and narratives—conveyed by a highly ambitious media and leisure industry that works in this attractive market—make it a fascinating landscape and a desirable destination for visitors from all over the world. Book titles of adventure and longing literature point the way, playing with the magical attraction of the Himalayas and the highest mountain in the world in particular. Longing for the Himalayas has found its way into everyday discourse through the expedition and travel reports of mountaineers. If the flat everyday life has leveled our lives, there is a way out into the heights. Then the well-trained and physically strong alpinist—and in his wake the proverbial *Maier*—is drawn to the great Himalayas. But ever since the foxtrot hit from 1926, people have been asking: what is he looking for there?[4]

> *The Myth of Shambhala*
>
> *An old Tibetan story tells of a young man, who set out on the road to Shambhala. After he had already crossed several mountains, he came to the cave of a hermit, who asked him: What is the goal that motivates you to cross these snow deserts?*
>
> *I want to find Shambala, the young man replied.*
>
> *Well, then you don't need to travel far, said the hermit.*
>
> *The Kingdom of Shambhala is in your own heart.*
>
> Edwin Bernbaum, *The Way to Shambhala.*

The myth of Shambhala stands out from the rich treasure of legends and hierophanies. The ancient texts speak of a kingdom hidden behind the snow mountains. There, a dynasty of enlightened kings guards the most secret esoteric Kalacakra teachings of Buddhism, preserving them for the coming *Kali yuga,* the evil era when truth has disappeared, destroyed by the greed for power and wealth and by war. The Tibetologist Edwin Bernbaum emphasizes the hardships of such a journey, described in the ancient texts. Shambhala can be reached only after an infinitely long and arduous march through desolate deserts and wild mountains. Many obstacles must be overcome to reach this distant sanctuary, and only a perfect yogi will be able to arrive there truly. This is indeed a beautiful and mysterious image—only through inner purification can we achieve outer perfection! The spiritual radiance can be experienced only by those who are willing to open their hearts and combine the wonder of a child with the wisdom of a fully matured person. Thus, as travellers, we can also be seekers, pilgrims who want to achieve a state of inner freedom, freed from mental and emotional unrest but also from fears. This kingdom may be beyond our reach, but it nourishes the longing for a place of purity and flawlessness. This is an idea also found in the central texts of Taoism: that the hidden land embodies the ideal community; in perfect peace of mind, perfect bliss is found.

Shambhala, that hidden kingdom situated somewhere in Central Asia, has become a place of happiness on earth in fiction and popular culture. In James Hilton's 1930s novel *Lost Horizon,* it is Shangri La, located deep in Tibet. And today, with the trivialization of its spiritual background and its reduction to advertising messages in travel catalogues, the search for that hidden kingdom is brought within the immediate reach of tourists. However, as Bernbaum points out, that fabled kingdom symbolizes that part of our world that remains hidden from our perception. As long as we cling to the illusions of our ego, we lack the necessary awareness to perceive the world truly. The veil of prejudice obstructs our view, and we cannot perceive the true world. Thus, modern-day interest in Shambhala reflects our longing for a more immediate experience of the true world.

This myth points to a universal form of something deeper, a longing for happiness as a state of contentment that exists in many cultures. We must all find our own Shambhala—that place, person, or idea which has the inspiring power to lead us through the inner journey to greater awareness and freedom. This inner journey is not a retreat from the world, but rather an attempt to experience the ancient teachings in order to recognize things in their real being.

The Shortest Way to Oneself Leads Around the World

In cultural studies and philosophical texts, even in the West, reference is made to this circumstance, and travel is interpreted as the experience of the world and as a way to oneself. In this departure into a heterotopia and at the same time to oneself, there is an element of utopia, as Klaus Kufeld[6] points out. The self-experience potential of travel is measured by one's openness in dealing with the uncertainty of human life circumstances. To extend all antennas and be ready to engage with the Other, the foreign—this is our chance to sharpen our perceptions and see new things, changing our perspective.

By contrast, the ready-made holiday package of today's happiness industry delivers foreign reality as merely a backdrop; more a state than a path, as our reality is merely relocated to another place. But even here, our individual perceptions are structured by their own systems of meaning and expectations, which form starting points for experiences. This process of perception, meaning-making, and experience construction draws on stored codes. Oral narratives, travel literature, images and imaginations are the sources for this construction process, with the help of which tourists create their worlds of experience. Karlheinz Wöhler writes that these ideas of people are happy spaces we want to experience because they are something out of the ordinary, perhaps even "places where the meaning of life unfolds".[7]

Happiness Seekers and Waste Separators

Many visitors to the Himalayan mountains are looking for the simple life. They want to return to the basics for at least a few weeks, to feel their own bodies and their limits, and immerse themselves in a spiritual space that the Tibetans call a *soulful landscape,* living in close connection with nature and connected to their gods or demons through rituals. The reality is that many tourists leave tons of civilization waste in the valleys of the Himalayas. It seems as if a power plant of unreason is working within us, encouraging contradiction. In the case of the World Natural Heritage Sagarmatha (Mount Everest) National Park, which has been entrusted to the protection of UNESCO and the entire international community, the need for careful use, cautious and gentle handling should be recognized as an absolute *sine qua non*. But the reality is that wear and tear and the destruction of biodiversity are accepted for short-term satisfaction, or out of

5 Utopias and Dystopias—Dream and Nightmare

convenience or negligence. One positive sign: for years, a basic waste management system has been implemented in Sagarmatha National Park, the entire region around Mount Everest, whose base camp had been considered the highest garbage dump in the world. This waste management system is intended to help restore dignity to the mountain, for in Tibetan Buddhism, the integrity of the landscape is associated with a life under the protection of the gods.[8]

The Austrian writer and Asia traveller Herbert Tichy (1912–1987), who hiked through the Himalayas several times and experienced the landscape "like a prayer", was deeply concerned with bringing the highest mountains in the world and their inhabitants closer to the people in Europe. He always wanted to convey the social context and his experiences in order to enable an understanding of the cultures, their traditions, and ways of life. In his book *Die Wandlung des Lotus* (The Transformation of the Lotus, 1951), he described this intention: "I would mislead if I only talked about the temples, the saints, and the Himalayas. They have been unchanged for centuries, one can only describe them as better or worse. The dramatic thing is the fate of the people who live in this region."[9]

Today, there is enough evidence of the magnitudinal changes in this region, not only concerning social and cultural upheavals, but also the habitat itself. However, this knowledge has not yet penetrated most Western circles. In the world of imagination, images of the conquest of the peaks dominate: it is a matter of human victory or defeat, of life and death. Or they are associated with H.H. the Dalai Lama or Buddhist monks and Indian yogis as symbols of spirituality. The myth of a place of eternal happiness in popular presentation resonates as a *standard narrative*.

The tightly woven connection of poverty and oppression, of war and ritual, of ethnic and religious diversity and conflicts, of wisdom and advancing destruction of the foundations of life, of the rapidly growing population and the increasingly dense settlement, which literally forces nature to its knees: these are realities that overwhelm us.[10]

The Himalayas as a tourism and experience space, as a manifest, multi-meaningful *other space* offers Western tourists a wide range of stimulants, cultural challenges, and situations in which they can test their travel competence and their intercultural tolerance potential. It can be both paradise and hell.

The Journey as Utopia

The analysis of the immediately perceived experiences, the evaluation of the knowledge in terms of its significance, and its integration into one's own intellectual cosmos require a further process of reflection. If tourists critically question their gain in experience and their own actions, a first step has been taken. They gain an idea of another world and thus of another way of being in the everyday world.

In such a view, the journey gains something utopian, as Klaus Kufeld explains with reference to Ernst Bloch in his *The Journey as Utopia*. The landscape journey becomes a knowledge motif, the tireless striving for deeper knowledge: *Homo sapiens* as a question, the world as an answer. However, to attribute this motif to tourism would be misleading, as only in cultural and educational trips is something like horizon expansion intended. Travel as knowledge-work is not what people seeking holiday relaxation strive for in their free time primarily.

For Kufeld, vacation offers a perfectly organized heterotopia, as a place of illusion that excludes everyday life, but

5 Utopias and Dystopias—Dream and Nightmare

a utopia that is only seemingly realized. He sees vacation as a travel pushed back to static, "more state than way", an illusion subtracted from the way, because the foreign reality is only understood as a backdrop and thus the difference to home no longer seems significant. The way there—to the holiday resort—becomes a distance again in this view, as once in the age of railway travel the in-between of starting point and endpoint was ultimately without meaning. If the journey is degraded to the goal, utopia falls out of the picture, because real utopia contains both there and back. Travel aimed at knowledge and knowledge as utopia means being on the way and staying on it, so it has no proper place. Touristic travel, on the other hand, is arriving travel, which has a place and only gives meaning to the distance or rather the way if it can assert visible experience character, because it either leads through an impressive landscape or can offer attractions such as a spectacular road layout. The way remains in memory, however, when for example a traffic accident, or a lengthy traffic jam experience has tarnished or destroyed the joy of vacation.

Vacation is most successful when it is not overwhelmed with too high expectations: the dream of freedom of mankind is not equated with two weeks on a Greek island or the area of a campsite assigned to the caravan trailer. Hans Magnus Enzensberger pointed out the contradictions of tourist travel in his *A Theory of Tourism* as early as 1958, before the beginning of the actual age of mass tourism. The more the bourgeois society closes, the more strenuously do we try to escape from it as tourists. Travel is one of the oldest and most common figures of human life. Tourism is based on this, but ultimately it is the serially produced form of temporary exit from the industrialized commodity world. Tourist travel has itself become a commodity and the industry has apparently found answers to some central needs in the form of more or less standardized offers.

Criticism has accompanied tourism since its beginnings. Ridicule is given to the tourist habitus—there is talk of 'sight-seeing hens', the high mountain peasants making fun of the city dwellers who wanted to climb the Alpine peaks, but did not know how to go about this. The term *idiot du voyage* made the rounds, ultimately resulting in a fundamental criticism of the phenomena of mass tourism, and the highly questionable ecological, cultural and social effects of the complete touristification of entire regions.

The writings of Swiss economist Jost Krippendorf (1938–2003) provided an empirically profound, comprehensible criticism of tourism and the circumstances that shape it. Back in the early 1980s, he called for a change in tourism, which he referred to in his social criticism as a *landscape eater*. At the same time, he convincingly demonstrated that the crisis of tourism is actually a crisis of everyday life in industrial society. In his *The Holiday People* (Die Ferienmacher menschen), he describes this cycle, according to which people go out to recharge their batteries, to restore their physical and mental powers. He speaks in the 'we'-form, because as much as we may criticize this context, we are all ultimately also a part of this system. We consume the climate, nature and landscape, culture and people in the areas we visit, which we have converted into therapy rooms for this purpose. Then we return home for a while—until the next vacation. But the desire to travel again and preferably even more often quickly arises, because life cannot be lived in a few holiday weeks and on a few weekends. The cart is overloaded, overcrowded with wishes and longings. This constant repetition of unfulfilled and unfulfillable needs gives the cycle its dynamics. We work in order to be able to take vacations, and we need vacations to be able to work again. We relax in order to be better harnessed prepared afterwards.[11]

The critical analysis in which Krippendorf described tourism as an escape from everyday life can be given even more empirical legitimacy in the context of today's accelerated society. Little wonder that many desperately seek healing spaces in the search for a work-life balance: the social fabric is becoming increasingly fragile, and many fear to fall by the wayside. Today, decades after the first edition of his book, we talk of limitless individual mobility—essentially based on fossil energy, highly deleterious to nature and the world climate. The criticism of the circumstances as well as of tourism itself has not lost any of its explosiveness. Krippendorf had also shown alternatives at the time: but little of this has found its way into the tourism of our days.

Dystopia Overtourism

For many people, the need to travel is a constitutive element of modern lifestyle. The tourism industry is responsible for fulfilling dreams—that is, visiting heterotopias in connection with some experience of the world—and has developed into a highly differentiated service industry with the increasing globalization of the travel market. Travel and vacation come together in the experiential place, which was previously the subject of utopian ideas. In places where many tourists come together, the mass tourism phenomenon of *overtourism* occurs, along with the fear that everyone will find what they are looking for at the same time, thereby destroying or impairing the experience. A fully booked hotel, a winter sports resort with crowded ski slopes, the overcrowded historic city centre or the beaches on the Costa del Sol and the Adriatic during the high season—all these places can convey such impressions. Overtourism refers to a state in which the number of tourists overstretches the local conditions: and this has a detrimental effect on travellers and locals alike. The limits of carrying-capacity and the resilience of a place or region are then reached or already exceeded. The local population's pro-tourism mindset shifts as the tourist experience becomes mixed with anger and frustration—and tourism develops negatively, towards the opposite of sustainability.[12]

Internationally, this topic has been widely debated, not least because of the extensive media coverage of the cries for help or protests of inhabitants of popular destinations such as Venice, Dubrovnik, Barcelona or Amsterdam. As these cities have been confronted with masses of tourists for years and are still on the bucket-list of many travel-hungry people, the protests have become an emotionally charged media topic. The city administrations have been forced to take political steps to keep the *revolt of the visited*—a term that was already circulating in the 1980s, referring to the massive protests of villagers in some developing countries against the tourist sell-out of their most beautiful beaches and landscapes[13]—within limits. And indeed, Amsterdam and Barcelona have made especially comprehensive efforts to conceptually cope with the excesses of tourism.[14]

In recent academic literature, there has been agreement that *overtourism* is a state in a tourism destination that corresponds to a perceived excess of tourism. How a critical threshold should be measured has accompanied the discussion about *carrying capacity*, the sustainability of a destination, for years. The indicators used for this are crucial. The extent of the tourist infrastructure and its utilization is one aspect, another is the traffic load or the use of mobility infrastructure. Thirdly, the perspective of the local population, whose opportunities for movement and development are affected, is included. Their mindset depends on the benefit they derive from the situation—or whether their living situation is negatively affected due to crowding, due to oppression and density experience, whether their tolerance towards the perceived acceptable changes reaches limits (tolerable rate of growth). Finally, visitor satisfaction is also a measurement indicator, the satisfaction of tourist desires, because endless queues and other unreasonableness, triggered by other tourists, can greatly impair the travel experience.[15]

The *Carrying Capacity Value Stretch Model* can serve for planning moderate tourism development, which determines a tolerance level with regard to an expectation level based on the current situation. If a red line is crossed compared to the current level, the local mood will develop negatively. Year-round tourist flows, overflowing rubbish bins, overfilled parking lots and coffee houses, large travel groups clogging the narrow streets—all these lead to unpleasant density experiences. If this becomes a permanent state, public discontent arises.[16]

Comprehensive tourism policy measures such as the creation of destination management, visitor guidance and new environmentally-friendly transport solutions are probably necessary to get the scope for managing large numbers of tourists. Once Corona pandemic abates, city tourism will pick up speed again, and museums and galleries will attract travellers from all over the world. The beaches and ski slopes have not lost their appeal; many tour companies have also used the pandemic to invest in improvements, in new lifts and other infrastructure. And the advertising machinery is running at full speed again.

Characteristic of the phenomenon of overtourism are the increasing affordability of long-distance travel for Asians or the middle class of emerging countries, cheap airfares due to inadequately taxed aviation fuel or the offers of so-called low-cost carriers, the—sometimes illegal—rental of living space to city tourists (Airbnb), which creates pressure on the housing market, the reckless behavior of tourists and the littering of public spaces and attractive places. In addition, tourism triggers higher price levels for locals, as well as challenging working conditions for service providers in hotels and restaurants due to seasonal concentration. Massive cruise ships and coaches cause the sudden appearance of too many tourists, overstretching the service providers and destinations.

From the perspective of travellers, the effects of mass tourism include large crowds, long waiting times, overpricing in restaurants and non-authentic cuisine, more increased traffic and noise, an oversupply of tourist services (souvenir and fast-food shops, kiosks) as well as rubbish and environmental pollution. Tourists find the excesses of their own kind very disturbing, as other tourists represent a conflict factor in their own tourist experience. Herd or horde tourism fundamentally impairs the individual tourist experience.

Waiting for hours for admission to the *Moisteiro des Jéronimos* (Monastery of St Jerome) in the Lisbon suburb of Belem does not enhance the monastic experience. Many cultural institutions have therefore introduced booking systems that require visitors to register in advance, after which they are assigned a specific time-slot for the visit. Patience and tolerance are particularly important for visiting the Tower of London, England's most viewed ticketed attraction, as well as for the Veronese balcony from which Juliet supposedly communicated love messages to her Romeo. Any case, visitors seeking uniqueness must simply accept that they share this experience with many others who are seeking the same thing with the same intention.

Endnotes

1. In my book *Auf der Suche nach dem Ort des ewigen Glücks* (Searching for the Place of Eternal Happiness: Culture, Tourism and Development), Kathmandu: Vajra 2014, I try to develop a minimal hypothesis for understanding

the mindset of the people and the central contexts in this region.
2. Kurt Luger & Martin Weichbold, Reisemotive und Reiseerfahrungen von Himalaya-Touristen (Travel motives and experiences of Himalaya tourists), in: Kurt Luger, Christian Baumgartner & Karlheinz Wöhler (eds.), *Ferntourismus wohin?* (Long-distance tourism—where to? The global tourism conquers the horizon), Innsbruck: StudienVerlag 2004, 395–416.
3. Eva Dietrich, *Himmlische Paradise* (Heavenly Paradises—Bhutan and the Himalayan region as salvation suppliers for the West), in: *Neue Zürcher Zeitung*, 17 June 2010.
4. https://de.wikipedia.org/wiki/Was_macht_der_Maier_am_Himalaya%3F, 01.07.2021.
5. Edwin Bernbaum, *The Way to Shambala*, New York: Anchor Press 1980.
6. Klaus Kufeld, *Reise als Utopie* (Travel as Utopia, Ethical and political aspects of the travel motif), Munich: Fink 2010.
7. Karlheinz Wöhler, *Touristifizierung von Räumen* (Touristification of spaces), Wiesbaden: VS-Verlag 2011, 9.
8. Project presentation and report at http://www.savingmounteverest.org/https://www.sagarmathanext.com/, August 30, 2021. For further activities see https://www.sagarmathanext.com/
9. The new edition was published by Edition Sonnenaufgang, Vienna 2016.
10. Jack Ives & Bruno Messerli, *Mountains of the World—A Global Priority*. New York: Parthenon Publishing 1997; Jack Ives, *Himalayan Perceptions, Environmental change and the well-being of mountain people*. London: Routledge 2004; ICIMOD-International Centre for Integrated Mountain Development, *The Hindu Kush-Himalaya Assessment*. Mountains, Climate Change, Sustainability and People, Kathmandu 2019. Online: https://doi.org/10.1007/978-3-319-92288-1.

11. Jost Krippendorf, *Die Ferienmenschen* (The Holiday Makers People, For a new understanding of travel and leisure), Munich: edition dtv, updated 1986, 15.
12. The term overtourism circulated quickly in the mass media because in some places there were citizen uprisings and attacks on tourists, to which the police had to respond with steering measures. See also Andreas Kagermeier, *Overtourism*, Konstanz: UTB 2021; Harold Goodwin: *The Challenge of Overtourism*. Responsible Tourism Partnership, Working Paper 4. 2017; online: http://haroldgoodwin.info/pubs/-RTP'WP4Overtourism01'2017.pdf; McKinsey & Company and World Travel & Tourism Council: *Coping with Success*. Managing overcrowding in tourism destinations, 2017, www.McKinsey.com, 02.06.2021.
13. This led to a tourism-critical movement that is still active today, involving various NGOs, not only in German-speaking countries. See Ludmilla Tüting & Jost Krippendorf, *Tourismus mit Einsicht* (Tourism with Insight), Starnberg: Working Group Tourismus mit Einsicht 1989; https://www.fairunterwegs.org/magazin/news/detail/1986-aufstand-der-bereisten/
14. Greg Richards & Lenia Marques, *Creating Synergies between cultural policy and tourism for permanent and temporary citizens*. Committee on Culture of United Cities and Local Governments, Rotterdam 2018; http://www.agenda21culture.net, 04.03.2021.
15. Harald Pechlaner, Christian Eckert & Natalie Olbrich, Ein zuviel an Tourismus? (Too much tourism? Status quo and solutions), in: *Tourismus Wissen*—quarterly, October 2018, 291–297.
16. Yoel Mansfeld & Aliza Jonas, Evaluating the socio-cultural carrying capacity of rural tourism communities. A "value stretch" approach, in: *Tijdschrift voor Economische en Sociale Geografie*, Vol. 97, No 5, pp. 583–601. https://doi.org/10.1111/j.1467-9663.2006.00365.x.

6

The Past Has Never Been as Beautiful as Today

Go to the museum! Connect with the world!
This is best done, when one understands the past.
Without knowing one's roots, one cannot understand oneself.
Vivienne Westwood.

I'm on vacation. No museums.
Leon de Winter, *Leo Kaplan*.

Tourism in historical ensembles corresponds to a time travel that leads "to the remains of great pasts", writes Valentin Groebner in his book *Retroland*. It is complemented by landscape sensations and the search for the beautiful. In doing so, one finds something soothing for mind and body. The journey into "time as a place of heightened sensation" allows travellers to immerse themselves in history without their own affiliation to modernity being threatened or questioned. The own origins or those of other societies appear foreign and enticingly exotic at the same time: attractive and sufficiently contrasting to everyday life to guarantee high intensity of experience. But they also establish a connection to the experiences of youth and trigger sensations, awaken feelings akin to melancholy.[1]

Heritagefication—Heredification: Cultural Heritage is Made

The past as cultural heritage, *lieux de mémoire,* as collective or as cultural memory—the entire area of memory cultures not only occupies historians and cultural studies, but also the tourism industry. With memory tourism, a wide field opens up for *cultural tourism,* a range of possibilities for returning to historical spaces. Visiting an historical site, tourists enter a space that is subject to a different chronology, full of past events that can be located, interpreted and remembered. A social framework is reconstructed that brings together history and generational memory, collective memory and personal memory in the sense of a configuration for identity formation.

Cultural artifacts of past times and generations such as buildings, monuments and memorials, events, works of art, rites and ways of life or historical figures convey cultural memory. They keep memory alive or continue it into the future beyond the present. Places or practices are heredified, made into cultural heritage by society or institutions like UNESCO. According to Aleida Assmann, the tourist preparation of such memory-places in the sense of UNESCO World Heritage corresponds to a new form of cultural memory.[2] It makes highlights of cultural creation globally easily accessible and takes on a mediating role. However, what is remembered with the World Heritage and which cultural phenomena of past realities enter cultural memory also depends on the present. Past experiences involve new re-interpretations of realities: contexts become experienceable and comprehensible in the present time, also repeatable and thus viable.

The 'memory power of things' resolves the dialectic of forgetting and remembering as well as of destroying and preserving insofar as a tourist use of the inherited—Karlheinz Wöhler[3] calls this "heritagefication", Christoph Kirchengast[4] "heredification"—at the same time means their protection.

In this way, tourism presents itself as socially and environmentally compatible—without, however, questioning more closely whether the presented and consumer-released memorial offer actually corresponds to the lived space, because memory processes are always deeply socially shaped. Cultural tourism thus becomes an attractive means of shaping history and at the same time a powerful medium.

Memory tourism focuses on places of importance for cultural memory. These can be places where the glorified past is central: places of remembrance of combat actions, war cemeteries, places where great natural disasters occurred, through which visitors can come to terms with human tragedies. The visit to memorials, such as Robben Island, where Nelson Mandela was confined for ten years, and which today houses a museum that presents the history of apartheid and South Africa, is referred to as memorial tourism. Like the Auschwitz concentration camp, it has been declared a UNESCO World Heritage Site. A significant memorial site which should not be missing in any Vienna city tour is the Square of Heros (Heldenplatz). Since the monarchy, it has been the scene of political stagings, "preservation depot" of Austrian history.

The difficult handling of history is particularly evident in the many countries affected by colonialism. An example of this are the churches and convents of Goa, the former capital of the Portuguese colonial empire. They were declared a World Heritage Site in 1986, at the request of the Archaeological Survey of India. The interplay of European architecture, native art, and craftsmanship created a unique ensemble of universal value. The Portuguese sailed around the world for Christianity and for spices; they ruled for centuries, proselytizing and subjugating the local population, and exploiting their resources. The culture of remembrance of Indian society—accentuated by the influences of Hindu nationalism—is therefore characterized by a *dissonant heritage*. However, European visitors today are told more about the Goa of the Golden Age than about the history of oppression: in today's tourism, the post-colonial countries welcome the successor generations of their former masters, the sahibs, with open arms.[5]

Splendor as well as misery are conveyed by the historic old towns like Salzburg, which are much-visited UNESCO World Heritage Sites. With the Prince-Archbishop Wolf Dietrich von Raitenau, the outwardly splendid epoch of the Salzburg Baroque princes began in 1587; and in 1803 the spiritual rule ended with Hieronymus Graf Colloredo. In-between, authoritarian rulers had dozens of churches built and the city decorated, with two turbulent centuries of absolute rule involving the expulsion of Protestants. Famines, floods, and hurricanes plagued the city; a rockfall killed hundreds of residents; the number of homeless and unwell increased steadily, and the marginal population was oppressed, persecuted, and expelled. The last witch execution took place in 1750. However, this era also saw the founding of the university and the first performance of an opera north of the Alps, Monteverdi's *Orfeo*. Fischer von Erlach's magnificent churches and palace buildings—the highlight of many city tours for tourists from all over the world—mark the end of the architectural history of the Baroque city.

Salzburg's Old Town is architecturally as unique as it is vulnerable: as a protected ensemble, it cannot tolerate too many interventions and architectural changes. It represents something incomparable, non-exchangeable, which therefore has no price-tag in today's world. Just as today's day workers from Nepal, who built the stadiums for the 2022 World Cup in Qatar in the desert, creating tourism infrastructure for starvation wages, they had to risk their lives for sacred architecture. Citizens and farmers groaned under the enormous tax burden so that this beauty could be created and financed. What was once available only to the Prince-Archbishop and his entourage now belongs to all of us, a World Heritage Site for all of humanity. In fact, we owe this achievement to the French Revolution: until then, monuments and the preservation of architectural heritage were matter for the Church and the Crown or the nobility. But proud citizens also defined the grand architecture as the heritage of the people and the whole nation. Consequently, the community has since been responsible for its preservation, but also for its adequate perception, interpretation, and appreciation of these cultural achievements.

In fact, this beautiful ensemble was created on the backs of those people who worked for the wealthy Prince-Archbishops. Surely we should remember not only the names of those who built or mastered the divine laws of statics and staged beauty, but also the achievement of the many nameless workers who contributed their part through the work of their hands, so that today everyone, residents and visitors, can feel connected to this extraordinary heritage of humanity and its contexts.

Profane Pilgrimage and Quasi-Sacralization of Space

The highest distinction for a space of such extraordinary universal significance consists in its inclusion in the list of UNESCO World Heritage Sites. By being elevated to World Heritage status and the associated cultural significance, a profane place or an important space for a faith community becomes a "quasi-sacralized space" for all of humanity.

In World Heritage tourism, visitors consume, study, and experience the fundamental elements of a culture, often icons of national identity, or the extraordinary beauty of a landscape. With the term *cultural landscape*, the World Heritage Committee distinguishes those objects that represent a kind of community work involving nature and man, which geography science refers to as a *landscape created by human culture*.[6]

Experiencing such a place distinguished as a World Heritage Site gives visitors the opportunity to experience themselves as part of history, to see themselves as part of a larger whole, because they come into contact with timeless orders. Wöhler speaks of a "sacramental experience" because it reveals the transcendent. This expresses an extraordinary appreciation and respect for certain places, memorials, or natural monuments. The history of the human spirit, the manifestations of its artistry, and the grand spectacles of nature are "sacralized" in the cultural understanding of contemporary societies, thereby forming a contrast to the almost completely desacralized cosmos.[7] The profane becomes something "holy" through this assignment of meaning: the elevation to World Heritage leads to a "canonization of spaces". A place, a region, a site is declared the "heritage of all humanity" of timeless significance. Out of the diversity of cultural artifacts, a few specific ones are declared to be most memorable.

In this uniqueness, in the exceptional and universal value, also lies a great touristic potential. There is a deep longing for spiritual experience, and holistic experiences, the desire to feel in harmony with the world. World Heritage tourists thus embark on a kind of secular pilgrimage. Experiencing these sacred places with their own senses is the main reason why tourists travel thousands of miles and climb hundreds of steps.

Conflict of Objectives Between Cultural Heritage and Tourist Marketing

Since the adoption of the World Heritage Convention in 1972, up to the meeting of the World Heritage Committee in the summer of 2021 in Fuzhou (China), 1154 heritages of humanity from 167 states have been included in the UNESCO World Heritage List. The criteria for inclusion in this list concern the outstanding universal value, the uniqueness, the authenticity (historical authenticity), and the integrity (intactness) of the individual objects. The sites selected by the UNESCO World Heritage Committee must be masterpieces of human creative genius, display outstanding natural phenomena or areas of exceptional natural beauty and aesthetic importance, or bear testimony to a cultural tradition or an existing or vanished culture.

All indicators on this list also apply to properties of tourist products—exceptionality, beauty, exclusivity, and uniqueness refer to dicisive *stellar moments in humanity* and reflect qualities on which economically successful tourism is based. For the Italian journalist and social theorist Marco d'Eramo, however, the designation of a place as a World Heritage site simultaneously pronounces its death sentence, because the masses of "barbaric tourist hordes" inevitably lead to the destruction of these monuments. The World Heritage title is merely the "beautiful soul" of the tourism industry: it soothes the conscience, and makes it possible to accept the complete marketing and the tourist devastations in the name of preservation.[8]

Here d'Eramo addresses a central conflict potential. With Heritage—cultural heritage or natural World Heritage—a fragile, non-renewable resource is meant. It needs protection to preserve its exceptional character for future generations. The threat affects tangible and intangible treasures equally, but the tangible, overcrowded historical old towns, buildings or cultural landscapes are in the foreground of consideration. However, the most threatened are traditional ways of life and cultural practices in the developing societies of the Global South, because their diversity is dying a slow death in the age of globalization. Significant disturbances in the cultural fabric are also caused by tourism, whose uncontrolled development is a factor of threat or change.

This fundamental conflict of objectives lies in the fact that Heritage as a system is governed by the underlying principle of preservation and transmission of what is to be passed on from generation to generation. World Heritage refers to the largest possible reference group—all humanity—and is oriented towards the common good. By contrast, tourism in the capitalist global economy is governed by the underlying principle of profit-oriented utilisation or consumption of landscape and resources. It follows the late modern concept of mobile leisure, of satisfaction of individual needs, and experience-oriented appropriation of the world.

The tourist development of natural spaces and especially the example of historical old towns epitomize this conflict of interests. The quasi-sacralization of the historical substance of a city that is protected as a heritage of humanity imposes a kind of temporal state of exception over this treasure. The freezing of the architectural ensembles fuels the dispute between preservers and renewers, leading to tension-filled confrontations, accentuated by the tourist marketing of local history and culture. Historical cityscapes form an antithesis to the modern automobilized city, which adapts or submits to traffic. In our age of climate change, however, old towns can offer a forward-looking perspective on sustainable economic activity and coexistence, which to some extent resists the dictates of unrestricted mobility, but also economic utilitarian thinking and profit calculation.[9] Such old towns form *heterotopias* which allow contemplation as well as entertainment, or become living spaces that combine work and leisure within walking distance. Preservation of old towns thus also has a justification in social thinking, with importance far beyond the beautiful appearance of an authentic façade design or the touristically motivated preservation of an historical experience world.

However, with high-quality cultural tourism a reconciliation of the objectives or principles can be achieved. This requires both an intensive engagement with the World Heritage through a tourism that enables meaning and deep experience, and the implementation of a tourism policy oriented towards sustainability criteria and the preservation of heritage.

World Heritage as a Destination of Longing

Venice, Florence, Dubrovnik, Salzburg—in many highly attractive tourist World Heritage sites, a rather voyeuristic approach to humanity's cultural heritage is at the forefront, increasingly degenerating into mass tourism with the city tourism boom. However, the high quality of what is offered requires a different approach, one which can guarantee due attention and appreciation: From seeing to interpreting to opening up the world: this requires the sensitive mediation of the high value of the symbolic memory figures in the form of specific architecture or the cityscape: *heritage communication* instead of *heritage marketing*. A quick walk-through in a crowd on standard routes, combined with a fleeting consumption glance of images and the obligatory selfie, certainly does not achieve this, nor can it generate a proper emotional, experiential or identificatory experience. The standard two or three-hour tour through the seventeenth and eighteenth centuries—souvenir shopping included—cannot do justice to the unique ensemble and its historical significance in any way.

Cultural tourism has been *en vogue* for years and shows enormous growth, which is reflected in the statistics of the UNWTO. About 40% of international tourism is attributed to trips that include cultural components. Cultural tourism is understood as a culturally motivated trip, assuming a certain basic interest of travellers in culture. On the other hand, tourists spending two weeks on a Mediterranean beach are also drawn to the ruins of Carthage and Knossos. Tourists vacationing at one of the Salzkammergut lakes or an alpine wellness oasis can also visit the festival city of Salzburg. The culturally motivated visit to the Historic Old Town makes them incidental cultural tourists. The same applies to winter sports enthusiasts who come from their ski resort on a badweather day to the city to visit the Baroque architectural ensemble, combined with shopping and the feel-good ambiance of a coffee house. The segment of explicit as well as incidental cultural tourists is continuously growing, and has become central to the travel industry for economic reasons, with valorization of culture also in the sense of observable lifeworlds, ways of living and traditions.

Tourist Value Creation from Cultural Heritage

As evident from the statistics of the UN World Tourism Organization, culture and nature have been promising growth markets for years. The treasures of cultures and nature form the raw material for high-quality products. Without them, tourism would not have become one of the world's fastest growing industries.

This dynamic has positive as well as negative effects on World Heritage sites. The contradiction inherent in tourism has prompted the UNESCO World Heritage Center in Paris to pay greater attention to this topic in World Heritage management. Two crucial questions emerge: First: How many tourists can a World Heritage site tolerate without harming the quality of the experience or the facility itself? Second: How many visitors are needed in order to generate economic benefits for the stakeholders of the World Heritage and ensure its preservation?

The enormous tourist value creation will be illustrated in the following using the example of the World Heritage and festival city of Salzburg.[10]

Few tourist cities have an image like Salzburg, so closely associated with culture. Wolfgang Amadeus Mozart and the Salzburg Festival, founded in 1920, shape its reputation as the music capital of the world. In the popular genre, the Hollywood film *The Sound of Music*—a multiple Oscar and Golden Globe winner in 1966—underscored the city's music image and made the beauty of the city and the surrounding region famous worldwide. In what the famed writer Stefan Zweig once rapturously described as the "successful marriage of human creativity with the God-given, architecture of man and poetry of nature", the experts of UNESCO also saw an outstanding universal significance. The Historic Old Town of Salzburg as an ecclesiastical-secular residence city, in which the dramatic cityscape with its historical structure and numerous ecclesiastical and secular buildings have been preserved, is unique for its connection with the arts, especially music in the person of its most famous son, Wolfgang Amadeus Mozart. The old town, with its beautiful churches, palaces, squares and quiet back-yards, was built on the rich mineral resources of the Salzburg mountains. But without the arts, which add a new dimension to the life of the city through music and drama, it would be little more than an oversized backdrop and reflection of past greatness. Since December 1996, the historic city centre has been a protected World Heritage Site.

Culture, or the interlocking of culture and tourism, is evident in many ways. Salzburg is a centre of classical music and a leading destination for value-adding quality tourism. Its brand image in the entertainment industry is due to the location placement of The Sound of Music, the annual Advent singing, the Christmas market, and the popular music and homeland films of the 1960s. Culture in all its manifestations is its trademark, shaping the typical habitus of the city, which becomes a cultural experience in itself. Millions of tourists are drawn to Mozart's birthplace to experience the city of music and visit the house where he was born. The city's fame extends far beyond its small-town significance. Mozart, together with the festivals, is the most important image and advertising medium, a "piece of world culture".[11] Marketed and consumed, he has become part of popular and mass culture. His image adorns Salzburg's most famous souvenir: the ever-popular chocolate and marzipan Mozart ball.

6 The Past Has Never Been as Beautiful as Today

In Salzburg, the number of overnight stays and day tourists has roughly doubled over the last 20 years. According to the value-added study cited here, in 2018 some 1.8 million overnight guests caused around 3.1 million overnight stays, plus about 360,000 overnight stays in Airbnb accommodations. About 15,000 hotel beds (almost two-thirds of them in the 4- and 5-star category) are available, another 2000 are under construction. There is also accommodation in the surrounding municipalities, often used by bus tourists visiting the city for the day. Estimates suggest that the number of annual day visitors—who arrive in their own cars or in one of the approximately 30,000 coaches—is around seven to eight million. For a city with 150,000 inhabitants, this means a very high tourism density. In the years before the Corona pandemic, locals demanded a reduction in day tourism, and visitor guidance from city politics. The majority of the population feels that the tolerance level of strain has been exceeded.

On the other hand, the population is proud of Salzburg's World Heritage and convinced of its benefit for the city. It helps to preserve the architectural heritage and serves as a seal of quality for tourists—provided they know about it at all. What monetary value is created, directly through the World Heritage or through the other attractions of the cultural city, is difficult to say precisely. The tourism of the city of Salzburg generates an annual value of about one billion euros, a third of which is triggered by motives associated with the World Heritage. The value created by the Salzburg Festival and the Advent singing alone is beyond 200 million euros. Both events are not only among the city's major image carriers: they also generate significant economic effects that go far beyond direct tourist value creation. For the province of Salzburg with its approximately 30 million overnight stays, a value of around 4.5 billion euros is assumed. In comparison, tourism in the federal capital Vienna—whose historic centre has been a World Heritage site since 2001 and Schönbrunn Palace like Salzburg since 1996—generates a value of around 3.7 billion euros and guarantees about 90,000 jobs.

According to Statistik Austria, the total Austrian tourism expenditures of domestic and foreign travellers together amounted to around 42 billion euros in 2017. The resulting direct and indirect value-added effects of 32 billion euros contributed 8.7% to the gross domestic product, which makes clear why Austria is economically so heavily affected by the Corona pandemic. Starting with the lockdown in March 2020, foreign tourism came to an almost complete standstill for several months. In the 2021 summer season, holiday hotel revenues were around 25% below normal levels. City tourism suffered most from travel restrictions, with a significantly higher foreign element than leisure tourism. That being said, the winter season 2020/21 was a loss for Austrian winter sports hotels and the cable car industry, with sales losses of up to 90%.[12] Before the pandemic, 320,000 people worked in the tourism and leisure industry nationwide. In the summer of 2021, a third of them were still on short-time work and 45,000 were unemployed. According to a recent survey among employees of the tourism industry in Vienna, 38% want to change their profession or the industry. The shortage of qualified workers in tourism will thus become more acute and is seen as a major challenge.[13]

World Heritage as a central component of a tourism concept has many positive aspects. However, it requires alignment with quality standards; there is a great need for legal frameworks to prevent the destruction or misuse of heritage. In its Operational Guidelines UNESCO requires from the management of a World Heritage site visions, management plans and their controllable implementation, as well as clear ideas or strategies for a tourism development oriented towards sustainability criteria. Laws for protection, such as the Old Town Preservation Act in Salzburg, are considered a prerequisite for responsible handling of architectural heritage, but there is no guarantee that they will be followed. Old towns are coveted spaces for real estate investment and speculation.

In many World Heritage sites, it is evident that this legislative protection of heritage is not sufficient. States do not fulfil their international legal obligations which they have entered into by ratifying the World Heritage Convention. In Italy, for example, valuable testimonies of antiquity and the Renaissance are decaying, some have been saved only by the cultural sponsorship of famous fashion companies; in France several are for sale and in the United Kingdom they are to be divided into heritage cash cows and charity objects.[14] The preservation of humanity's heritage is very insufficient in many countries; in the case of war events, as recently in Syria, Iraq or Afghanistan, there has been partial or complete destruction of World Heritage sites. Such endangered World Heritage sites are on UNESCO's Red List, together with those whose preservation is not guaranteed due to lack of consideration in urban planning, such as the Austrian capital Vienna.[15]

Living Tradition with Experiential Character

Tourists want to experience something, to know the special features of the place, and the destination or what is visited and consumed there must trigger emotions. Performances of traditions, rites and folk festivals as part of a tourist product have to compete with the stagings of artificial experience worlds. An *experience*—that magical word of the tourism industry—becomes such when the viewers are captivated by an event and the situation triggers an emotionally charged feeling. An experience breaks with everyday life for a certain duration: it has a beginning and an end, and is highly subjectively evaluated. Here we may distinguish between various kinds of experience areas relevant for leisure or vacation: *biotic* experience (unusual physical stimuli); *explorative* experience (searching for information or exploring, trying out, curiosity about something special); *social experience*, seeking contact with others to compensate for social deficits in normal everyday life. And an *optimizing experience* occurs through the subsequent processing and narration about it.[16]

The promise of unique, once-in-a-lifetime experiences is the key success factor behind the boom in artificial leisure and experience worlds such as theme parks, urban entertainment centres with multiplex cinemas and shopping, fun and adventure pools or brand lands. At the *Swarovski Crystal Worlds* in Wattens, on the motorway between Munich and Innsbruck, there is even an "experience toilet". Thousands of small Swarovski crystals shine from the ceiling, sparkling like stars in the dimmed atmosphere. The experience spiral intensifies the experience through an increasingly differentiated range of offers. Activism, emotionalisation, a focus on sensations and extraordinary events—in short, *Opus* instead of *Passus*—are the hallmarks of today's successful entertainment industry.

Increasingly at the center of tourist attention is the presentation of ritual actions in the form of customs and festivities in the annual cycle, ciphers of cultural heritage. The advancing globalization has given increased appreciation to the local or one's own environment. Cultural heritage has been defined as revaluing the obsolete, endangered, discarded or extinct.[17] Through this and through forms of staging, a second life is breathed into the affected things, places or practices. Christoph Kirchengast[18] explains this using the example of the hay barns found in many alpine regions of Austria, which also adorn postcards and tourism brochures. Their original function—the storage of hay—has long been lost, often due to structural changes in agriculture (keywords: mechanisation, silage bales). Nevertheless, these barns are often actively maintained, sometimes even with public funding. Today, they are symbols of the summery, intact cultural landscape in the Austrian Alps and thus serve tourism: they have an economic value.

If cultural heritage were only the protection of the endangered, the discarded, then preserving it from oblivion would not do justice to tradition and time-honoured ways of life. Cultural heritage also includes techniques of craftsmanship or ways of life in the context of specific environmental conditions, and it concerns both town and country equally. Cultural heritage is therefore not reducible to representations that serve a specific tourist purpose by marketing particularly picturesque or attractive aspects of a cultural context. The typical also contains the general, but the question of the origin of a particular tradition does not answer the question of the quality or usefulness of certain cultural practices in meeting today's challenges.

Perception, emotionality and memory are essential components in the construction of tourist experiences. Tourists travel by following images, imaginary geographies acquired through socialisation, myths and narratives that become desires and expectations. This process of constructing experience and meaning is based on stored codes. In tourism, guests are thus presented with a 'play'. Austrians, Bavarians, South Tyroleans and Swiss perform on the front stage with stereotypical Alpine images aimed at foreign guests.

For tourists, oral narratives, travel literature, photographs, etc. are sources for this construction process, in which they create their own worlds of experience. From the history, cultural backgrounds, myths of a destination, brand-forming themes can be created, which through professional storytelling emotionally charge the tourist space. Aspects of tourism and stage are combined in the concept of staged authenticity: *performed authenticity*. The logic of action of the theatrical gesture is oriented towards skilled performance: the ritual becomes a performance. The line between the genuine, the meaningful in life, and the stage version is thin, as is the task of bringing past, present, and future into a balanced relationship. The sensitive and responsible handling of cultural heritage poses a massive challenge indeed.

City dwellers, with their disdain for the provincial and folkloric, contrast with the mood of the population in peripheral regions. However, when they travel as tourists to the mountains or the pre-alpine lake landscape, they often perceive the culture experienced on site as an attraction and are captivated by the magic of authenticity.

Real or not Real?

Hardly any region can do without its own cultural summer festival these days. Just as the dances of the Serengeti Maasai or the Himalayan shamans, the rituals in the touristified Alpine villages are sheer exoticism for guests from China, Philadelphia—or Westphalia. Corpus Christi procession, the stallion and cow drive, finger hooking and Alpine *Hundstoa ranggeln* (a type of wrestling listed as Austrian intangible heritage)—they are all eye and ear experiences! Folklorists like Konrad Köstlin ask themselves whether it might not be thanks to tourism that one or the other custom still exists, because the colorful folk cultural customs would hardly make sense without tourism. "Local self-celebration needs its audience, needs resonance. Tourism and folk culture belong together, they are twins."[19]

Well, life in the countryside has changed and so have the ways of life. Many farmers have become tourist service providers because they can no longer live off their field yields. Tourism today means *experience space management*, which includes landscape and people and has effects on their everyday life and culture. In the Salzburg region, the population is adept at integrating cultural heritage and tourism into functioning livelihood strategies. And indeed, if planned and operated with care, this can become a lucrative form of economy based on local resources and conditions. As such, it influences everyday life, shapes forms of life and rituals of interaction. The positive link between modernity or appropriate keeping up with the times and locality is reflected in approaches to integral regional policy. Regional culture thus does not necessarily lead to a brand identity consisting only of historical set pieces and memory quotes: it can become a lived speciality, which contributes to the attractiveness of the region, and a unique selling point for the destination.

The question is not whether a custom or ritual appears real or unreal, but whether and how traditional cultural forms and techniques can be integrated as innovations into one's own way of life and thus contribute to the unrepeatable shaping of a place. Every holiday village has the right to develop a stage form and present it to tourists with folklore heritage evenings. In the long term, however, a region will only be able to convey its appearance, its peculiarity and its uniqueness in a credible way if it links the current living environment with the historical depth of its regional culture. This can only be achieved if the local population, together with the nature they have shaped and their contemporary culture, are not presented exclusively as a retro musical, but are allowed to experience and explore themselves in their complex and diverse worldliness.

Tourism professionals would do well to recognise and celebrate local culture in the design of their tourism products, rather than inflating every little hut party into a centuries-old tradition or artificially inventing new ones. The trivial Tyrolean evenings in Austrian alpine valleys have long since served their purpose. Guests want entertainment and an easily consumed form of fun, to join in the drinking and dancing, but it has to be sustainable, the aura of the performance has to be credible. And indeed, the list of living traditions in Austria, recognised and listed by UNESCO as intangible cultural heritage, is particularly impressive because the alpine topography has produced a particularly rich diversity over the centuries.[20]

Tradition today no longer means the "sacred transmission of truth" or "belief in the inviolability of what has always been".[21] Living traditions are those conventions, instructions for action, knowledge in dealing with nature: they are craft techniques—but also performing arts and social practices that build on old knowledge with answers and solutions for current challenges.

Rituals such as carrying the tall Samson figures in the Salzburg Lungau or the Glöckler runs and Krampus passes not only have a high community-supporting significance for the local population, but also relevance for tourism.[22] Festivals are so attractive because they colorfully and diversely accentuate the playful aspect of the ritual.

Dramatic elements and strong emotions are expressed. Since celebrations, in their exuberance, do not usually place high demands on authenticity or origin, they can be relatively easily placed in different contexts. This makes them cheap raw material for tourism in the sense of an unrestrained culture of experience. While mass tourism is reminiscent of crude medieval festivals, cultural and educational travel broadens the horizon. If the encounter with the other culture is successful, it can even lead to a different view of the world. Tourism itself is characterised by ritual processes, from the welcoming rituals (the inevitable "Schnapserl", a glass of home-distilled schnapps, or alphorn blowing) to the post-processing of travel impressions at home (sorting and labelling photos, notebook entries). Like the ritual celebrations, it participates in the transformation, the play, the pleasure that contrasts with the seriousness of life. The holiday destination becomes a counterspace that compensates for the defeats of everyday life, a place of happiness where the utopia of the good life becomes reality.

Driven by the emission-free driving force of human curiosity, tourists flirt with the view behind the stage, into the privacy of the visited and their way of life. They are, first and foremost, voyeurs who usually only chase after what they already know from pictures, such as sights or worlds of experience that have been prepared for tourists according to a script. Tourists are looking for exactly these snippets of reality in the form of an accessible and therefore consumable offer that has authentic features, does not look too much like off-the-peg, prêt-à-porter, and allows direct sensual contact. To this day, Viennese holidaymakers dress up in the Styrian Ausseerland costume before setting off for the Salzkammergut. This makes the experience 'unique', worthy of being ennobled by souvenir photos, coveted souvenir trophies.

The claim to authenticity of any performance remains questionable in an age of fake news, commodification and imitation. What we think of as folk culture was in fact the way of life of peasants and craftsmen in the villages, towns and suburbs of the pre-industrial age. This diverse culture only gained socio-political relevance when it was discovered by the aristocracy and bourgeoisie, who spent their romantic summer holidays in the provinces wearing dirndls, loden jackets and lederhosen. What they brought with them to the city was politically purified, aesthetically refined and stylised, creating a second culture—a construct whose authority-critical and resistance potential was largely eliminated. Instead, picturesque elements such as costumes, rituals, folk songs, house forms, folk medicine and folk piety were emphasised. Through folklorisation, this folk culture gained prestige and increasing importance in politics, the media and, of course, tourism. Because elements of folk culture were considered threatened, they were protected from the effects of industrialisation and urbanisation, collected, preserved, maintained in associations, museums and archives, and coated with a varnish of authenticity. At the same time, this was the framework for the formation of nation states and regions whose ideology emphasised the rural-agricultural heritage.[23] Folk culture gained considerable importance in self-identification during this time. Even today, many regional cultures are characterized by this, although this rural world increasingly disappeared in the twentieth century.

Folk culture plays a central role in the shaping of a distinct popular culture and thus in the construction of homeland and identity. In the stylization of traditional costumes as badges of local folk affiliation, of alpine flowers as national symbols of the natural, of certain architectural forms as recognition signs of grounded construction like the pseudo-rural saddle roof, of show customs and folk music, the ethnologist Bernhard Tschofen sees a program with recognizable ciphers that led to an identification-establishing emblematics of an ideal design of Austria images.[24] These were picked up by the expanding culture

6 The Past Has Never Been as Beautiful as Today

industry after the Second World War, when the Heimat (homeland) became a tourist lure. In the genre of the homeland film, a reprogramming of collective memory took place. Central to these "homeland makers"[25] was the beautiful landscape as a backdrop for a peasant world as city people wanted to see it: harsh, patriarchal, but also idealized as a summer resort. In the countryside—so it was suggested—the old world was still (somewhat) in order. The cinematic myths of village life contain numerous cheap identity offers, although the reality of this rural world had long since fallen into a frenzy of modernization. The peasant everyday culture mutated into folklore, the mountain landscape became a consumable object. In these films, people took a holiday from history, repressed the traumas of the Nazi era, and at the same time advertised for vacations in Austria.[26]

Nothing of substance grows in the Alps above 1500 m above sea level—except tourism. For peripheral Alpine valleys, it has become the most important, if not the only, thriving economic sector. The prerequisite for this is a well-functioning service infrastructure of people and machines, but also the use of local culture and resources. In customs, parades and dress codes, the presentable tradition is celebrated and business is done. Both locals and visitors know that the content of the old folk songs, the traditional costume bands, the wood carvings, most of what is called 'folk culture', does not correspond to today's reality and is often served up especially for the paying guest. Tourists consider those hotels that can be ascribed to the "Cowshed Gothic" to be typically Tyrolean, even though they do not reflect the traditional Tyrolean building style in architecture or painting.[27] For tourism, certain elements of the local, regional, Austrian, etc. are absolutized as typical or as a special feature and stylized into easily recognizable signs. Around them, products are designed, the success of which is decided by the guests. The quality of a coherent product is therefore measured in a process of interaction with the target groups. This is most evident in gastronomy, where regional cuisine has established itself as a territorial speciality with an authenticity that goes beyond geographical indications of origin.

However arduous and alien the rural world may seem to visitors, tourism only presents rural culture in its Sunday best, in a romantic or nostalgic context. After all, the everyday life of the locals is not an attractive product. It is not their normality that interests visitors, but the colourfulness of their cultural forms, their peculiarities, their life based on nature and animals, their exoticism—and all this in a choreographed form. Sometimes tourism creates its own tradition. In 1996, the Salzburger Bauernherbst (Farmers' Autumn) was introduced as a marketing measure to enliven the low season. And indeed, the number of overnight stays by tourists in the low season has increased by 20% since then. Organised by the Salzburg Provincial Tourist Office, which is responsible for tourism marketing, the communities were invited to offer events for themselves and their guests under the motto "Celebrate, Taste, Culture". Locals celebrate together, sit together, make music, dance, cook and have fun—with tourists as guests and as festival visitors. It is a successful advertising campaign for regional products without any claim to authenticity. Here and there, village and religious festivals merge with tourist offerings—on the verge of commercial folklore and sometimes beyond.[28] Similar concepts can be found throughout the Alpine region, all advertising their regional specialties.

With the production of homeland symbols, homeland films, homeland calendars, homeland literature, Austrian picture books and school texts, from homeland 'schmaltz' singers to the Grand Prix of folk music, from tourism advertising to the souvenir industry, which—as Bernhard Tschofen writes—rests on the three pillars of *edelweiss, gentian* and *alpine rose*, popular culture provides a multifaceted iconography of the Alps as a true land of hearts' desire. It constructs the image of the Alps at home and abroad and forms the experience software for those seeking recreation and adventure. In doing so, it helps to make offers as target-group-specific as possible and to fill the beds of post-modern tourism.

Reference to the Present as a Programme

The years when museums stood with their backs to the wall, when they were considered hopelessly outdated because the winds of modernity blew over the traditional, and these storage rooms of the old stood in the way of the promise that everything can only get better: those years are now over. Sociological analyses point to a trend of *retrospection*—which in the language of tourism marketing is called *regrounding*—and thus also an appreciation of the traditional, in which one seeks security and a kind of grounding, a cultural anchoring, and a foundation of enduring societal values.

At the same time, we find not only in the Alpine countries that time-honored societal practices or cultural heritage, cannot be exclusively associated with a backward-looking political ideology. Especially in Germany and Austria, folk culture, seen as an expression of territorial identity, was ruthlessly exploited for political purposes, but here and there it also marched in the front line. In recent decades, mass-market folk culture has experienced an unprecedented appropriation and trivialization, a trimming down to what is marketable in the media, in popular films, beer tents and music halls. Tourism, in turn, has incorporated the folklorized and beautified appearance of festive culture and various show customs as aesthetic embellishment of its offer.

The concept of *tradition* is undergoing a long overdue semantic redefinition, thanks not least to the efforts of UNESCO. With the World Heritage Convention (1972), the Convention for the Safeguarding of the Intangible Cultural Heritage (2003) and the Convention on the Protection and Promotion of the Diversity of Cultural Expressions (2005), it has initiated a socio-political discourse of global dimensions, which is linked to the rest of UNESCO's cultural program and takes on a specific form in the context of the Alps and tourism.

This discourse has drawn attention to the vitality of a wide range of traditions, rescuing them from the obscurity of dusty storerooms of the meaningless and the supposedly hopelessly outdated. It is probably no coincidence that this is happening at the dawn of the twenty-first century, as the age of fossil fuel draws to a close, the unquestioning belief in ever-increasing prosperity erodes, and creative thinking about new visions invites a look in the rear-view mirror. Many an unfashionable or submerged cultural asset is now resurfacing and becoming part of our everyday culture. Complementary medicine, with its reliance on traditional healing methods, is a direct expression of this new openness. This can also be seen in fashion. Suddenly, dirndl dresses and lederhosen are being worn—here and there with a wink, but also with a new self-confidence. Traditional costume, with its high quality workmanship and materials, is being greatly appreciated as a piece of clothing with substance.

Museums are places of memory, archives of collective memory—but not only that. Their tasks go far beyond the preservation and interpretation of the past. If one defines cultural heritage as living tradition, then the path taken by museums into the present also becomes logical in terms of a programmatic approach. Beyond the task of preservation, the contemporary discourse has long since stepped in, the commitment to present the cultural practices in which society finds its forms, but also to locate them in their contexts, to contrast their former significance with current contexts. This affects the entire range of intangible cultural heritage, the diversity of life practices, everyday and artistic expressions, knowledge and skills that communities understand as part of their cultural heritage.

Media such as smartphones and social media services not only provide convenience and ad hoc information, but can also multiply the appeal and experience. Memory institutions such as museums use these digital capabilities to make connections and relationships visible and easier to understand. Augmented reality applications combine the real world with artificial realities and can transform ideas into powerful images.

Not only tourists, but also locals often realize only through systematic processing that these practices are passed on from generation to generation, but also reinterpreted and designed. Through this self-assurance, personal reference is created, through continuity also belonging and identity, to varying degrees even togetherness. Museums thus become unique places of societal self-understanding and platforms for public discourse, through exhibitions and information programs also initiators of a critical public based on scientific debate.

The relative newcomer in the European museum landscape, the EcoMuseum, is taking exactly this path. Such a museum is meant to present characteristic natural and cultural goods in a holistic way with the participation of the population and the inclusion of the landscape, in order to preserve them within the framework of historical heritage—but also to establish a connection with the present. This is a way of making ecology, ethnology and history or cultural change in a regional context accessible and experienceable as a living tradition.

The reinterpretation of places and practices as cultural heritage triggers processes and (self) dynamics that would not otherwise occur. The culture of revival is widespread: in fashion, outdated things are brought back into circulation as vintage fashion; the retro wave is often used to develop new value creations from the properties of cultural heritage. It seems as if a tidal wave of memory has swept across the world, confronting the digital in an analogue way and gaining a foothold in social discourse as a culture of memory. Tourism is a clear example.

For the Austrian anthropologist Christoph Kirchengast, the 'musealization' debate that has been going on since the 1980s is closely related to the discourse on cultural heritage. *Musealization* refers to the increasingly common practice of preserving the past in the present. An often-heard explanation for these trends is that they are a kind of compensatory, collective search for identity, responding to general anxieties about the disappearance of the familiar. These fears are fuelled by a perceived shrinking of the present: the process by which the validity of scientific knowledge, everyday technologies, or fashion seems to be diminishing—nothing seems to last. A second cause of the memory boom lies in the experience-orientation and the sensory method of seeing that are ingrained in contemporary postmodern society.[29]

Along with the term 'musealization of the world', we could equally speak of a 'heredization' of the world and our everyday lives. While heredification refers to the concrete making of heritage, heredization refers to the global trend, the societal inclination towards different forms of collective heritage. This term makes sense, as the concept of cultural heritage has gained significant momentum in recent decades. The number of UNESCO World Heritage Sites is now considerable, and there is no end in sight to this trend. The UNESCO Convention for the Safeguarding of the Intangible Cultural Heritage has given further impetus to the discourse on cultural heritage. Living traditions, customs, folk culture, practices in dealing with nature—they are given a new and greater significance. The suitability for everyday use can be seen as a key success factor of the *cultural heritage concept*, as all these things or practices can be brought from the past into our present experience and incorporated into popular culture and tourism.

The ongoing process of modernization of life contexts has undoubtedly led to many solution mechanisms losing their significance and some traditions having little or no relevance today. But culture must be understood as a process. Because society has to deal with innovation and face new challenges, the value and order system must also change. New cultural practices need to be developed or existing ones recognized and suitably adapted. The 2003 UNESCO Convention for the Safeguarding of the Intangible Cultural Heritage is a counterweight to the process of change, often triggered by technical innovation. This is more than a call for order, it is a chance to become aware of the achievements and accomplishments of cultural practices and to understand tradition as the guardian spirit of the new in an innovative context.

Today, modernity has dislocated society from the certainties of tradition and set other identities in motion. Beyond material prosperity, people are increasingly asking questions about inner meaning; they are looking for points of reference in the familiar and the local; they are seeking subjective security in our age of accelerated technological, economic and social change. The sociologist Heinz Bude claims that people are fundamentally conservative, "because they see no sense in a world that is falling apart, in which everyone wants only to protect themselves and their own".[30] Hope without optimism is the mood of the future. It appeals to needs that are apparently best satisfied on holiday, or that are perceived as the window of opportunity for successful loss compensation. At the same time, a counterspace is sought that compensates for the defeats suffered in everyday life. Thus, cultural heritage not only tells stories from the past and immerses past times through a sepia filter: cultural heritage offers a new perspective on the world. It combines old knowledge with new meanings, helping us to find new self-confidence.

Endnotes

1. Valentin Groebner, *Retroland. Geschichtstourismus und die Sehnsucht nach dem Authentischen* (History Tourism and the Longing for the Authentic), Frankfurt: Fischer 2018.
2. Alaida Assmann, Das Welterbe als neue Form des kulturellen Gedächtnisses (World Heritage as a new form of cultural memory), in: Kurt Luger & Christoph Ferch (eds.), *Die bedrohte Stadt* (The threatened city: Strategies for human-friendly building in Salzburg), Innsbruck: StudienVerlag 2014, 19–25.
3. Karlheinz Wöhler, Heritagefication: Zur Vergegenwärtigung des Kulturerbes (Heritagefication: On the presentification of cultural heritage, in: Kurt Luger & Karlheinz Wöhler (eds.), *Welterbe und Tourismus* (World Heritage and Tourism: Protecting and Utilizing from a Sustainability Perspective), Innsbruck: StudienVerlag 2008, 43–58.
4. Christoph Kirchengast, Einverleibtes Vermächtnis (Incorporated Legacy), in: Kurt Luger/Karlheinz Wöhler (eds.), *Kulturelles Erbe und Tourismus* (Cultural Heritage

and Tourism—Rituals, Traditions, Stagings), Innsbruck: StudienVerlag 2010, 301–325.

5. Kurt Luger & Mihir Nayak, World Heritage: Sacramental Experience, Heterotopia, and Sustainable Tourism, in: Kurt Luger & Matthias Ripp (eds.), *World Heritage, Place Making and Sustainable Tourism: Towards Integrative Approaches in Heritage Management*, Innsbruck: StudienVerlag 2021, 69–86.

6. For further details about the World Heritage Convention, nomination to World Heritage, and the mandate to preserve, see website of the UNESCO World Heritage Committee, https://whc.unesco.org/.

7. The philosopher and sociologist of religion Mircea Eliade dealt with the two existential situations of the sacred and the profane in *Das Heilige und das Profane*, Frankfurt: Insel 1998.

8. Marco D'Aramo, Die Welt Im Selfie (The World in a Selfie), Berlin: Suhrkamp 2018, 111.

9. I have illustrated this conflict in the context of property speculation and investors' calculation using the example of Salzburg, which is typical for the various interests of stakeholders in a World Heritage city. See Räume des Begehrens und des Bewahrens—Welterbe und Stadtentwicklung im postdemokratischen Zeitalter (Spaces of desire and preservation. World Heritage and urban development in the post-democratic age), in: Thomas Herdin & Franz Rest (eds.), *Kurt Luger: Medien, Kultur, Tourismus* (Media, Culture, Tourism, Transcultural findings about worldview and lifeworld), Baden-Baden: Nomos 2018, 253–278.

10. https://www.researchgate.net/profile/Kurt-Luger/publication/337914712_Studie_zum_messbaren_Wert_des_Welterbes/links/5df29f04a6fdcc28371d1a79/Studie-zum-messbaren-Wert-des-Welterbes.pdf

11. Leopold Sedar Senghor, *Österreich als Ausdruck der Weltkultur* (Austria as an expression of world culture), speech at the opening of the Salzburg Festival 1977,

Official program of the Salzburg Festival 1977, Salzburg 1977, 27–39.
12. https://www.statistik.at/web_de/statistiken/wirtschaft/tourismus/index.html
13. https://www.derstandard.at/story/2000127688879/hotellerie-umfrage-38-prozent-wollen-branche-wechseln, 21.07.2021.
14. Ron van Oers, The Economic Feasibility of Heritage Preservation, in: Wiliam Logan, Mairead Nic Craith, Ullrich Kockel (eds.): *A Companion to Heritage Studies,* Chichester: Wiley-Blackwell 2015. https://doi.org/https://doi.org/10.1002/9781118486634.ch22.
15. https://wien.orf.at/stories/3002904/; https://www.unesco.at/kultur/kulturgueterschutz/rote-liste-des-gefaehrdeten-erbes-der-welt/, both 01.08.2021.
16. Walter Kiefl & Reinhard Bachleitner, *Lexikon zur Tourismussoziologie* (Dictionary of tourism sociology), Munich: Profil 2005.
17. Barbara Kirshenblatt-Gimblett, *Destination Culture—Tourism, Museums, and Heritage,* Berkeley: University of California Press 1998.
18. See note 4.
19. Konrad Köstlin, Reisen, regionale Kultur und die Moderne (Travel, regional culture and modernity), in: Burk-hard Pöttler & Ulrike Kammerhofer-Aggermann (eds), *Tourismus und Regionalkultur (Tourism and Regional Culture)* Vienna: Verein für Volkskunde 1994, 11–24, here 13.
20. www.unesco.at/kultur/immaterielles-kulturerbe/
21. Karsten Dittmann, *Tradition und Verfahren* (Tradition and Procedure, Philosophical Investigations into the Connection of Cultural Transmission and Communicative Morality), Norderstedt: BoD 2004.
22. Matthäus Rest & Gertraud Seiser (eds.), *Wild und Schön* (Wild and Beautiful, The Krampus in Salzburg Land), Vienna: LIT 2016.

23. Reinhard Johler, Volkskultur(en) (Folk Culture(s)—European with a future?), in: Yearbook of the Austrian Folk Song Work, Volume 57/59, Vienna: Verlag der Provinz 2009, 59–70.
24. Bernhard Tschofen, Berg—Kultur—Moderne (Mountain—Culture—Modernity), Vienna: Verlag für Gesellschaftskritik 1999.
25. Gertraud Steiner, *Die Heimatmacher* (The Homeland Makers), Vienna: Verlag für Gesellschaftskritik 1987.
26. Kurt Luger, Tradition—Ritual—Inszenierung (Tradition—Ritual—Presentation: Cultural heritage between conservation and tourist appropriation), in: Kurt Luger & Karlheinz Wöhler, (eds.) *Kulturelles Erbe und Tourismus* (Cultural Heritage and Tourism), Innsbruck: StudienVerlag 2010, 15–45.
27. Bernd Schmidt, *Tourismus-Architektur im "Alpinen Stil" als touristische Botschaft* (Tourism Architecture in the "Alpine Style" as a Tourist Message. The "Lederhosen Style" as a Cultural Symbol), Diss. Phil., Salzburg 1998.
28. https://www.salzburgerland.com/de/bauernherbst/ accessed 1.8.2021. For a general picture see Franz Rest, Urlaub am Bauernhof (Farmhouse Tourism), in: Kurt Luger & Karlheinz Wöhler, (eds.) *Kulturelles Erbe und Tourismus* (Cultural Heritage and Tourism), Innsbruck: StudienVerlag 2010, 145–157.
29. See note 4.
30. Heinz Bude, *Das Gefühl der Welt—Über die Macht von Stimmungen* (The Feeling of the World—On the Power of Moods), Munich: Hanser 2016, 127.

7

Alpine Tourism: Fair Weather Zone in Climate Change

… the German Tourists.
At first they did not come for a long stay. They just spent a few nights in one of the few hostels to climb some Alpine peaks. But they talked a lot and told many stories. They bragged and, using their homeland as an example, they constantly made suggestions on how the village could be better organised and changed. Even today, Germans are considered the ideal tourists in this area. They are easy to please because they do not have high demands as to comfort and hospitality. They only need two things: generous portions of food and several newspapers. Although the German tourists were looking for the pleasures of nature, they brought an urban atmosphere with them. They urbanised the village.
They told stories and glorified their climbing tours.
Soon even the peasant boys began to climb the mountains. Skis, which had already been used during the war, were gradually re-imported by holiday-hungry city dwellers. Soon the village youngsters were making their own wooden skis.

Lucie Varga, *Ein Tal in Vorarlberg* (A Valley in Vorarlberg—between the day before yesterday and today, 1936)

The Alps, that region where Europe touches the sky, have been a place of longing for centuries. Surrounded by myths, they have always offered plenty of room for human imagination and the urge to push the horizon. Once the great mountain range was more of an obstacle and the steeply rising rock formations sent a cold shiver down the backs of their visitors, but with modernity came the penetration of this space, as inaccessible as it was enticing. Gradually, the Alps began to be used in multiple ways: they provide a habitat for the rural people who farm the valleys further up; they became a source of industrial and economic development, as miners and workers extract resources from their depths and process them in local factories; they are a retreat and sanctuary for fauna and flora—but, above all, they become a favourite recreational and experiential space for their visitors, the tourists. As a tourism habitat, the Alps are a place of anticipated happiness, an emotionally charged geography for travellers, whose numbers now run into the millions.

The history of Alpine travel is also a history of mobility in the age of individual delimitation, especially of automobility. Through railways, roads, paths, cable cars, hotels, and a network of mountain refuge huts, the Alps were made accessible, becoming a place of temptation and desire for city dwellers.[1] The variety of monumental landscapes, rock formations, flora and fauna, and distinct cultures have been drawing scientists to the mountains for about 250 years. The vertical challenge tempts alpinists to dare the impossible, to climb the highest and steepest face and report the success back home. On somewhat less risky terrain, mountain hikers move, interacting with the elements and dealing with their own physical limitations—with many in search of the meaning of life with minimal luggage. All these endeavors have in common the eternal fascination that stems from the ideal image of a mountain landscape as a new *world landscape*, where we can experience being part of a larger whole.

The Alps consist of spaces where survival and progress could be secured only through hard work. But they also host many places of the good life—now increasingly threatened by climate change, unbridled infrastructural development, and the deliberate destruction of the ecological foundations. As an illusion industry with a claim to fulfillment, tourism is a beneficiary of the beauty of this space—but it must also make its contribution to preservation within the framework of a sustainable regional development perspective. In the long term, development planning will have to aim for a strategy that can be understood as *preserving and protecting progress*. This includes the preservation of the natural spaces, the careful use of existing resources, and a moderate tourist infrastructure that enables profitable value creation.[2]

Over the past hundred years, tourism has contributed significantly to cultural and social change in the Alps, accelerating the transformation of rural culture towards a service society. But it was above all the major transformations of the global economy and the socio-political and agricultural policy decisions of the European Union that assigned a new role to the people of the Alps. The agricultural contribution of the inner Alpine regions gradually dwindled to a minimum; and without secondary income, mountain farmers found it difficult to make ends meet.[3] Like few other industries, tourism in peripheral areas secures the livelihoods of the locals, creates decentralized jobs, and thus also preserves endangered Alpine ways of life.

Fascination Mountain and Valley

In the history of tourism, mountaineers play a key role. The first ascents of the Alpine peaks, the founding of the Alpine clubs, and the excursions into the untouched and elemental: all these embody the romantic ideology of tourism. The romantic dream of leaving one's own restrictive civilization behind to gain freedom fuels alpinism to this day. The pushing of boundaries and overcoming of barriers through personal physical performance continues to fascinate, and the values of today's performance society provide the driving force in the sense of renewable energy. Mountains are spaces of longing and resonance, with which we can build a relationship; similar to music, they can trigger pain and joy, despair and feelings of happiness, enabling an intensity of self-perception.

Tourism in the Alps has shown steady growth for many years, especially in the core tourist areas and where the mountains are infrastructurally developed and easily accessible. The reasons for the high appreciation of the Alps as a holiday and leisure region can be found in the industrial economic model of society, which creates an externally determined working day for many, with its stressors. Indeed, the development of productive forces brought prosperity, and scientific-technical progress became the driving force of a growth-oriented competitive economy. However, the lives of individuals have been subjected to increasingly rigid temporal rules, and these dominate life processes. Not everyone succeeds in combining the multitude of challenges, action and experience episodes into a coherent life under these circumstances. Many people see alpine tourism as an opportunity for correction and authentic experience, to reduce accumulated mental and physical fatigue. This favours the collective social desire towards nature experience and culminates in the need to do something for one's health, to have control over time and to move in free spaces, to enjoy peace and to be able to experience silence. The lifestyle of health and sustainability (LOHAS) has developed from a niche to a growing trend that influences many forms of tourist opportunities, offering a varied leisure culture with new outdoor sports which includes almost all social groups. Given these social context conditions, the growth of alpine tourism will continue in the foreseeable future—and, inevitably, some alpine regions will reach their limits or exceed their carrying capacity.

The Conquest of Height

Alpine tourism owes much of its rise to the lure of the mountains. In the late 1800s, popular lookout mountains such as Rigi (1871) and Pilatus (1889) in Switzerland were the first to be developed with cogwheel railways and summit hotels. The cable car from Bolzano to Kohlern, which began operating in 1908, is the oldest aerial cableway in the Alps for scheduled passenger transport. Since 1912, the Jungfrau Railway, an electric cogwheel railway, has been running from Grindelwald up to the 3454 m high station on the Jungfraujoch in the Bernese Alps. In Austria and Bavaria, a dozen cable cars—like those to the Rax in the Vienna Alps, the Schmittenhöhe near Zell am See, the Patscherkofel near Innsbruck, the Hahnenkamm near Kitzbühel and the Predigtstuhl near Bad Reichenhall—became operative between 1926 and 1937. The technical development of the Alps, initially criticised by the Alpine clubs, was eventually accepted as an extension of the tourist possibilities. In this respect, alpine tourism is a true child of modernity but also of postmodernity: today it drives the development of alpine spaces, their economization and their functional adaptation to residents and visitors alike.[4]

The tourist conquest of height, the view from above as an expression of superiority, the mental elevation above others, the possession of the panoramic totality, the aesthetics of summit experiences, the experience of the body and the overcoming of boundaries— a hundred years earlier, all this had provided reading and discussion in the alpine magazines. These were not yet predominantly mouthpieces of the sports and leisure industry, but saw themselves as educational and cultural media, as important actors in the discourse on modernization and preservation of tradition.

The economic and socio-political debate about the forms of use—in addition to tourism, this particularly affects road construction and the water or electricity industry—was finally replaced by increasing awareness of the need to protect this culturally significant topography. Also the image of the Alps changed, as reflected in today's brochures and websites where ski lifts, cable cars, ski slopes and water reservoirs are marked, and biker routes, forest roads and summit restaurants are indicated.

But despite the thousands of kilometres of motorways, ski slopes and cross-country ski trails that crisscross the mountains, the cable cars and golf courses with their innumerable holes, the vast leisure infrastructure of hotels and spas, the widespread image of the Alps today is only partly true. The romantic postcard image contrasts with the claim that they are actually 191,000 square kilometres of gradually overused mountains. The Alps: Europe's gymnastic apparatus, a giant playground and amusement park?

Indeed, alpine geographical studies by Werner Bätzing and others nuance this picture: 40% of Alpine communities have virtually no tourism at all, with 46% of all tourist beds concentrated in five percent of the communities.

Half of the hotel infrastructure is accounted for by only 300 of the communities, including large agglomerations like Chamonix. Tourism is spatially concentrated in a few areas, but people all over the Alps are affected by traffic to and from holiday destinations, especially by transit traffic, such as on the Brenner route and the Tauern motorway, to an almost unacceptable extent. There are about 600 real tourist communities: those with a mono-structured tourism industry without significant out-commuters, which have often developed from high-altitude farming villages; ten percent of all Alpine communities. Only eight percent of the population live there and the actual tourism area is only some 10,000 sq km.

These tourist communities, often grouped together as hiking and skiing areas, consists of about 300 skiing areas. Approximately five percent of the total Alpine area is used for tourism and some 10–12% of jobs are in Alpine tourism, making it an important economic sector, but not the dominant one.[5]

Spatial concentration of Alpine Tourism

However, the Alps are one of the largest and most important tourist regions in the world. The effects of tourism extend far beyond the steep slopes and well-maintained hiking trails. Werner Bätzing emphasizes that due to the flourishing tourism business, the Alps can by no means be described as a disadvantaged region. Many villages in Tyrolean, Salzburg or Vorarlberg valleys, in Valais, Engadine, South Tyrol or Val d'Isère rank high in per capita income, although there are significant regional differences. With the advent of industrialization and the general loss of importance of alpine agriculture, a process of ecological, economic and social neglect or degradation of large parts of the Piedmont and Ligurian Alps began, but also of individual areas in the French Southwest Alps. In Slovenia, even in the fully 'air-conditioned' Switzerland and also in Austria, there are some areas that offer no survival opportunities for small and medium-sized businesses far beyond the pastoral idyll: they are economically and socially desolate, and already largely exploited ecologically.

Today about 60% of the Alpine population live in towns and conurbations, partly on the edge of the Alps, mostly in favoured valley locations. Two thirds of jobs are located there, and the development of municipalities since 1870 shows that the Alpine population is increasingly becoming urban. Not only do they live in greater numbers in the conurbations, but they are also becoming city dwellers in their thinking. This development is not only due to tourism, but also to the modernisation of society in general, to the greater permeability of lifestyles and to the narrowing of the cultural gap between town and country. Young people in particular do not want to do without urban amenities. The educated elite is the driving force behind cultural and social change, bringing with it new ways of thinking and living. However, not so many remain in the countryside because there are few jobs that match their education.[6]

Around 13 million people inhabit the approx. 190,000 sq km large Alpine arc, in 6200 communities and eight states. In terms of possible permanent settlement area, the Alps are thus one of the most densely populated regions in the world. Due to its diverse effects on local economies, on the social fabric and cultural heritage, on the environment due to emissions caused by leisure traffic or through the consumption of landscape and raw materials—simply put, due to the enormous ecological footprint—tourism in the Alps deserves the highest attention.

With an estimated 200 million tourists visiting the Alps each year, tourism has long since overtaken agriculture in terms of value added. In many Alpine valleys, local communities are entirely dependent on tourism, which is increasingly unevenly distributed across the region. This is also true for Austria, which, with about 28.5% of the Alpine area, shows a spatial concentration of tourism in relatively few municipalities. According to a 2016 study, the top 20 Austrian winter sports resorts with their cable cars generate more transport revenue from winter tourism than the remaining 298 resorts combined, and with 20 million overnight stays account for around 46% of the total. Economic success obviously requires extensive tourism infrastructure, especially accommodation and cable cars, and large structures benefit from this.[7]

There is a widespread but erroneous impression that tourism has taken over the whole of the Alps. Today there are about 9.9 million tourist beds in the Alps, including 1.3 million beds in hotels, 3.2 million beds in parahotels (commercially rented second homes, alpine club huts, etc.) and an estimated 5.4 million beds in privately used second homes. In his extensive studies, Werner Bätzing has shown that tourism is highly concentrated in the Swiss, French and Italian Alps, while it is more decentralised in the Bavarian, Austrian, South Tyrolean and Slovenian Alps. But even in the Austrian Alps, the number of large tourist centres with more than 5,000 beds increased significantly between 1985 and 2014.

Concentration processes also affect the ski areas. At the beginning of 2016, there were 634 spatially contiguous ski areas across the Alps with a total of 26,515 km of slopes. Austria has the most ski areas in terms of numbers, but many are smaller, while France has significantly fewer, but very large ski areas. In terms of the length of the ski slopes to the respective alpine area, then the Swiss and French Alps are the areas most intensively developed for downhill skiing.[8]

A key benchmark, a competitive advantage in the competition for ski tourists in times of climate change, is the altitude of the ski resort. 45% of all ski areas do not exceed 2000 m in altitude and are therefore heavily threatened by the effects of climate change. Only 12 ski areas in the Alps reach an altitude of 3,300 m. Of these, nine are in the Western Alps and three in the Eastern Alps. In addition, 46 glaciers developed for skiing offer possibilities for skiing even in warm winters.

The ski areas are located in the territory of 817 municipalities or 13% of all Alpine municipalities, with these municipalities usually concentrated in the core of the Alps, near the main Alpine ridge between the Cottian Alps and the Lower Tauern. These are usually municipalities with a below-average population and an above-average municipal area. At the top of all political units is the federal state of Salzburg, in which 47 of the 101 Alpine municipalities have at least one ski area, followed by Tyrol, Valais, Vorarlberg, and South Tyrol. This means that even in the hotspots of ski tourism, less than half of the municipalities have ski areas. The total area corresponding to the immediate ecological impact zone of a ski area through slopes, reservoirs, buildings, etc., covers about 5,600 km^2—or 2.9% of the Alpine area.[9]

Feel the Heartbeat of the Alps

Where does Alpine tourism stand today and where will it develop, given the massive environmental changes already underway? Since the *Year of the Mountains 2002*, when the UN drew attention to the vulnerability of mountain areas as part of a year-long campaign, several trends can be observed, not all of which can be seen as Alpine-wide preservation strategies. A detailed overview and status report are provided by the 35 contributions in the book *Alpenreisen* (Alpine Travels—Experience, Spatial Transformation, Imagination), edited by Kurt Luger and Franz Rest (see note 1 below).

The most striking feature is certainly the advanced economization and intensification of tourist infrastructure, which promotes a spatial concentration of Alpine tourism. The marketing is done through spectacular stagings and events in tourism centers, but also through the construction of viewing platforms, skywalks, and suspension bridges in summit regions and cable-car tourist investments for experiencing the glacier world (such as Kitzsteinhorn, Zugspitze, Dachstein). This is accompanied by cross-media tourism marketing, presenting a leisure alpine dream landscape of ideal images. All this leads to the elevation of the Alps or the Alpine experience through technical means, with landscape or nature in the tourist centers increasingly pushed into the form of collective leisure and amusement parks and completely subjected to economic logic.

This goes hand in hand with a diversification of outdoor sports and a revaluation of the Alpine summer, previously a weaker season in the shadow of winter. Speedhiking, e-biking, Hike & Bike, fixed rope routes, potholing, llama or Haflinger horse trekking, ultralight mountaineering, flying fox and canyoning on the one hand, and moderate alpinism, hiking, forest bathing, spiritual hiking and guided adventure tours for young and old on the other— the mountains offer something for everyone. The tourism industry provides a rich and varied range of activities to satisfy as many of man's outdoor needs as possible.

Another long-standing trend with cultural overtones is linked to the general desire to spend time outdoors, to experience alpine summers and regional cuisine in upmarket hotels. It is encouraged by the cultural accentuation of the local, the emphasis on cultural heritage, and can thus be interpreted as a counter-movement to long-distance tourism. Even before the Corona pandemic and the restrictions on overseas and long-distance travel, holidays in one's own country were very popular. Summer retreats or farmhouse holidays are typical of the boom in holidays in the mid-altitude regions of Bavaria, South Tyrol and Austria. In times of uncertainty and the increasingly irritating waiting times at overcrowded airports, the comfort zone of home is also getting closer emotionally, and the relatively high standard of accommodation and gastronomy in Alpine tourism is being appreciated more.

The wellness, spa and health tourism sectors have seen particularly strong growth. Health-promoting holidays at high altitude, combined in health resorts with ever more generous wellness landscapes and luxurious spa facilities, are increasingly replacing stays previously authorised in health institutions and paid for by social insurance organisations. The extensive privatisation of the social security system, whereby the individual now has to pay for the maintenance of his or her own physical strength, has led to significant growth in the feel-good economy.[10] In the Austrian holiday hotel industry, the wellness component is fully implemented by around 1100 hotels with 120,000 beds. The low-interest fiscal policy of recent decades has also enabled companies with little equity to invest in expensive (often exaggerated) SPA worlds. This has also made the providers less affected by lengthy periods of rainy weather. In 2019, over 20 million overnight stays—about 15% of all 150 million overnight stays—were attributed to this segment.[11]

The dynamic development of this sector in the upper price segment, a huge investment in high-quality building materials and technical equipment, in design and comfort, has inspired alpine architecture to ambitious projects, new hotel constructions and creative solutions at cable car stations, ski jumps, museums and other public buildings. Architecture and aesthetics in the alpine landscape have become a controversial topic, especially in the context of questionable investor projects that have been sprouting up like mushrooms on many prime spots in the Alps. Chalet villages, apartment hotels or second homes have been the gold of the Alps are considered the "concrete gold" of the Alps—a highly problematic development that has not yet been brought under control by spatial planning laws.

This dangerous circular development was accentuated by the Corona pandemic. The demand for second homes in a safe environment, with the simultaneous possibility of working from home, has increased significantly. In fact, it all began in the 1950s with the decline of agriculture and traditional crafts, both of which were increasingly losing their base to industrial production. In the decades that followed, many farmers were forced to sell their land; others were able to adapt as artisans, getting new work from investors building apartment blocks and second homes. But what was considered sustainable proved increasingly problematic, and the recent building boom has been accompanied by a wave of public criticism about the 'selling out' of the homeland. The model, which only works if there is a constant flow of new orders, is now working against the interests of the local population. The more second homes are built, the more do land prices rise: in some regions they have long been unaffordable for locals. These second homes are usually used only for a few months each year, but the municipality has to maintain the infrastructure—water, sewage, electricity, roads, snow removal, etc.—causing year-round costs. The more is built, the more is agriculture pushed back. Increasingly depleted villages gradually lose their typical character. Younger, more active residents are increasingly drawn to the cities or to other places where the action is.

The negative ecological consequences are also considerable. The consumption of landscape leads to further sealing of areas and to the sprawl of the landscape, sociologically speaking to a 'suburbanization'. In the process, the traditional rural cultural landscapes with their pronounced small-scale structure and their great ecological diversity disappear.[12] They are replaced by banal and uniform settlement areas, causing the landscape to lose its small-scale structure, its biodiversity, identity and attractiveness—precisely those attributes for which city dwellers actually seek out peripheral areas. If the spiral continues in this way, it destroys the cultural and ecological prerequisites for sustainable regional development.[13]

In Switzerland, this threat was recognized earlier and a strict limitation for second homes was introduced[14] by referendum in 2012. The Austrian spatial planning laws—different from province to province—offer too many loopholes and thus little legal means against this loss of cultural landscape and quality of life.

The trend towards the car is also unbroken. A large part of the national traffic volume in the Alpine countries is due to the leisure and tourism industry. Leisure mobility, a central achievement of current industrial societies, is primarily defined as car mobility. Very popular are the drives over the Alpine passes—Gotthard, Dolomites or over the Col de la Bonette, the highest Alpine road in Europe on the border to the Parc National du Mercantour. The Grossglockner High Alpine Road, a typical tourist road located on the outer zone of the Hohe Tauern National Park, is visited by around one million tourists each year and is one of the most popular attractions in Austria.[15] More than two thirds of Austrians travel by car for their summer holidays and to winter sports resorts. Even more so for the Germans, the car is by far the most important holiday vehicle. With increasing travel intensity, a cross-border mobility management will be necessary in the future to avoid long-term environmental damage and to keep the burdens from the increased traffic volume for travellers and those visited within acceptable limits.

However, all forms of tourism that partially or completely forego the use of cars also experienced a noticeable upswing, which can be summarized as *gentle tourism*. This primarily concerns retreats or hiking and cycling tourism, journeys that lead to self-discovery, that serve to slow us down and provide relaxation through peace and a slow pace. A variant of this is nature-based tourism, which takes place mainly in protected areas, in national parks, Natura 2000 areas, etc. These all have in common the extensive renunciation of car mobility, using public transport to get there and locally provided electric vehicles, horse-drawn carriages or bicycles. The year 2017, proclaimed by the UN WTO as the *Year of Sustainable Tourism*, clearly promoted this trend.[16]

In stark contrast to this is the intensified expansion of ski resorts (hotels/chalets, gastronomy, garages/parking lots, shopping malls, adventure parks, roads) and the enormous investments of the cable car industry for development, replacement and improvement of lift facilities, as well as for the supply of artificial snowmaking technologies. This expansionist drive is also seen in the merging of ski areas into networks, as they want to set a benchmark in the competition by increasing the number of piste kilometers of skiing slopes. This has led to numerous disputes between the cable car industry and environmental organizations like the Alpine clubs in recent years.

Mountain Experience

> *The German is a very controlled person. You have to set him between 0.5 and 1 per mille alcohol, where the loss of control begins—then you can milk him.*
> Lois Hechenblaikner, Tyrolean photo artist (*Profil*, Austrian weekly magazine, May 30, 2021)

Do the mountains really have to reinvent themselves, as a headline in an Austrian daily newspaper suggested? The Alps are completely boring, the Tyrol is in deep sleep; thus the verdict. This seems strange, as the Tyrol is the region with the highest density of ski resorts worldwide—79 ski areas, 480 lift facilities, and 49% of all Austrian ski days are in this federal state. However, skiing has indeed reached its zenith; further growth is hardly possible, and the number of skiers is declining for demographic reasons. The core target groups are aging, and young people are not replacing them. Moreover, the cable car industry is subject to certain limits due to spatial planning requirements, legally prescribed environmental impact assessments (EIA) and resistance from civil society to new developments. Nowadays, cable car companies invest mainly in comfort and capacity increases, as the development potential is primarily seen in the diversification of the offer, in the variety of products. From pure nature to Snowfun Parks, everything is on offer: wellness, culinary delights, feel-good vacations on the one hand, more adrenaline on the other. This is especially problematic for Austria, whch is home to the largest hotel and ski industry in the Alps. As a small country, Austria is heavily dependent on foreign skiers—especially on the Germans—where some two-thirds of its approximately 52 million ski tourists (first-time entries) come from.

'Staging' is en vogue: many tourism professionals are convinced that guests want staged mountains and that this is the only way to create attractions. The art of staging mountain 'experiences' consists in creating a special atmosphere. The second magic word is 'authenticity'—emotion through real experience, encounter with indigenous people, with the secrets of nature, the confrontation with the immediacy of the given. With the landscape as an attractive backdrop, the search is for the symbiosis of high-quality staging and the true mountain experience.

Herein lies a contradiction: the dispute over the right balance between intensive and extensive forms of tourism, as it is called in the Tourism Protocol of the *Alpine Convention*, provides ample potential for conflict. The dispute is exacerbated by some cable car companies, which are expanding rapidly in several holiday regions and want to drive further expansion through new developments. The *International Commission for the Protection of the Alps* (CIPRA) has therefore called for a halt to the conversion of glacier landscapes into ski slopes and the extensive expansion of ski resorts throughout the Alps, arguing that the "ruinous competition" between ski resorts leads to the destruction of nature and the landscape without contributing to the development of a sustainable economy. The *Austrian Alpenverein* (Austrian Alpine Association), which has some 650,000 members, has long been calling for an Alpine spatial plan to consolidate the tourist offer and break the growth spiral. It should point out alternatives to technical tourism and formulate final limits to expansion in order to guarantee the preservation of natural areas as a complement to intensively used economic and tourist regions.

Across the Alpine region, citizens' initiatives and civil society organizations are seeking to counter numerous projects for the new development of glacier regions, the infrastructural upgrading of previously little-used natural spaces and the development of new areas for ski tourism. Applications to host the Olympic Winter Games in the Alps are hardly supported by the local population, due to the feared major construction sites and the additional traffic load.

However, leading representatives of the Austrian cable car industry consider this to be completely unjustified. The planned merger of two glacier ski-areas with partial new development of a glacier region in Tyrol ignited the socio-political dispute at the frontline of ecology and economy. The cable car industry turned against the Alpine Convention, a legally binding agreement signed in 1981 and ratified by the Alpine countries. The umbrella organization CIPRA was accused of being less interested in protection than in preventing projects and "in depopulating the valleys so that lynx, bear, and wolf can be settled".[17]

The positions are indeed far apart and are justified by completely different interests. While some demand visible investments in sustainability measures and a preserving development in view of the climate crisis and UN Agenda 2030 for Sustainable Development, the economic interests of others push for further expansion of the already highly touristified mountain regions. In the 2010s, there were numerous spectacular winter sports mega-projects and mergers of ski areas in various parts of the Alps. The largest contiguous ski area in Austria, with 305 km of slopes and 87 ski lifts, was created by merging the major ski resorts around the Arlberg. In France, the *Compagnie des Alpes* controls 12 of the most important French ski areas, operates fun parks, and also holds shares in noteworthy Swiss and Italian ski stations. As the largest corporation in global winter tourism, the Compagnie sells approximately as many ski passes as the entire European competition combined. With further concentration, there is likely to be only a small number of large ski associations in a few years. This market consolidation will be driven by economic considerations on the one hand and the effects of climate change on the other.

Challenge Climate Change

Replacement and comfort investments in winter sports resorts have so far guaranteed a flourishing business. The number of skiers has been constant to slightly declining for years, but this has less to do with snow conditions. Technical snowmaking ensures that up to 80% of the slopes are largely snow-sure, depending on the region, even at medium or even lower altitudes. Artificial snow has become an indispensable basis for winter tourism. In Austria alone, there are over 420 storage basins that supply around 25,000 snowmaking systems with water and, if it is cold enough, produce sprayed snow with an energy cost of 4.2 kWh per skier. In the state of Salzburg alone, 85% of the ski slopes are snowed, which corresponds to 4700 ha of slope area or 0.65% of the state area. About six billion liters of water are stored in 120 storage lakes, which are taken annually from the natural water cycle, which corresponds to about half the consumption of the city of Salzburg. Approximately 50 million euros are invested annually in snowmaking and the expansion of necessary facilities. This is accompanied by the consumption of an estimated 24,000 megawatts of electricity—a considerable effort for a 'real nature experience' in winter.[18]

However, it must also be noted that in recent years some cable car companies have started sustainability processes, and produce renewable energy on their own to reduce their ecological footprint. The economic results in winter sports or ski tourism were satisfactory to excellent—until the pandemic.

Despite the stagnating number of skiers, the Austrian cable car industry is the driving force in the expansion of alpine tourism. At around € 1.5 billion, cable car revenues already surpass those from gastronomy. Calculations show a total expenditure generated by the cable car industry, with its almost 3000 lift facilities of about € 8 billion annually and a derived value added of € 4.3 billion. Prior to the pandemic, the industry recorded around 600 million transports. With these impressive figures, representatives of the cable car industry appear confident and underline their high importance for the entire national economy of the country.[19]

Since 2001, this sector has been achieving annual investments averaging around € 500 million, thus investing more than a third, in some years up to half of the revenues. In the winter of 2016/17, the total investment volume rose to an estimated € 710 million. Of this, € 171 million were allocated to snowmaking systems. No wonder that the prices for lift tickets are also rising and skiing has become a relatively expensive leisure activity!

Some thriving resorts or cable car companies also face the problem that many destinations are struggling to survive. Low-lying or smaller resorts are almost always loss-making, some are on the verge of bankruptcy and need public subsidies. Larger lift companies in favorable tourist locations, on the other hand, can set aside two- or even three-digit million amounts for future investments or invest in corresponding comfort, safety, and snow supply techniques.

In the long term, however, the outlook for Alpine tourism does not look so rosy.[20] The *Forum for Climate and Global Change* of the Swiss Federal Government alarmed the industry with its forecasts in 2007, as it linked warming in mountain regions to a decrease in snow reliability. However, some studies of snow depths and temperature development in the Alps consider the extent of the feared effects of climate change to be exaggerated. *MeteoSwitzerland*, for example, in its 2015 survey on the past 50 years, concluded that high-alpine temperature development in winter is characterized by periodic warming and cooling phases, and that neither a clear warming nor a clear cooling can be proven for the high winter in Switzerland.[21]

The *Austrian Panel on Climate Change* calculated the snow reliability of Austrian ski areas based on the 100-day rule (three months of natural snow and over the Christmas holidays). According to this, their number decreases to 81% with one degree of warming, to 57% with two degrees, and to 18% with four degrees of warming. All climate-change studies predict more pronounced warming for the Alps than for lowland regions. This will be accompanied by glacier retreat, thawing permafrost, increased landslides, extreme precipitation and more frequent storms and floods. Thus, the potential for danger will increase and the costs of maintaining the infrastructure will rise significantly. Already, hardly any winter sports destination can manage without artificial snow, and a temperature increase of two degrees doubles the expenses for snowmaking. Fewer and fewer ski areas in the foothills will be able to operate profitably.[22]

The forecasts for summer look better, but the landscape will change, the dangers in the mountains will increase—hardly advantegous for cable-car companies seeking to compensate losses or declines in winter with a significant intensification of summer business. That this will probably be possible only through further development of Asian markets. However, some investments in the redesign of mountain stations indicate that urban forms of entertainment with great technical effort will increasingly degrade the natural mountain world to a backdrop.

Alpine Skiing—from Lilienfeld to Ischgl

The special position of skiing and winter tourism in Austria deserves its own consideration.[23] As early as in 1896, the Austrian skiing pioneer Mathias Zdarsky published the first textbook on alpine skiing under the title *Die Lilienfelder Skitechnik* (The Lilienfeld Skiing Technique). His technique was a further development of the practice prevalent in Scandinavian countries (Norway's Telemark County is considered the origin of skiing). Zdarsky laid the foundations for enjoyable skiing in alpine terrain as a mass sport as well as a competitive sport. While the Nordic practice was more aimed at movement in less-steep terrain or mastering obstacle courses or ski jumps, the focus in the Alps was on mastering steep terrain without falling. In the early days of skiing, this necessarily also included uphill movement, as the first mechanical lifts appeared only in the late 1920s. This also led to the separation into the *Nordic tradition* (cross-country skiing and ski-jumping) and the *Alpine style* of skiing, downhill or slalom as downhill only, because there were no more uphill sections to master in the competitions. In ski touring or ski hiking, which has been gaining more and more followers recently, this development seems to be coming back to its starting point.

With the invention of the *stem turn* by Hannes Schneider, the *parallel turn* by Toni Seelos, and the *wedeling technique* by Stefan Kruckenhauser, a specifically Austrian culture technique of movement in snowy alpine terrain developed, which was taught as the High School of Skiing in ski schools. During the second half of the twentieth century, it became the international standard, continuously further developed through material and movement-technical innovations. In addition to the experience of movement and the experience of nature, social relationships have remain a major attractor of alpine skiing. Club life, competitions in ski clubs, and the typically Austrian practice of *après-ski* enjoyment have become part of a leisure-oriented lifestyle—a *ski culture*. This has become a significant economic factor, as skiing is by far the most important sport of winter tourism. Austrian ski victories on Austrian skiboards—Fischer, Kneissl, Blizzard, Atomic were all Austrian companies almost in patriotic service. The brands still exist today, although the companies have long since been absorbed into international corporations. Skiing is the only sport in which Austria has global significance. In no other country are winter sports in all their forms as popular as in Austria, and especially alpine skiing is one of the most popular sports. In a self-designed event or group-experienced mix of excitement, enjoyment, effort, and relaxation, it also fits into the fitness and health philosophy of modern society; for many, it is a part of everyday and leisure culture.

Modern skiing in Austria has found a home in over 1200 clubs and ski associations, organised in regional associations and united in the Austrian Ski Association (*Österreichischer Schiverband, ÖSV*). While the promotion of popular sport is voluntary, competitive sport is clearly profit-driven, and the ÖSV controls several commercial companies for the marketing of skiing. Beyond the clubs—where volunteer ski instructors serve as instructors—skiing techniques are taught in local ski schools. Professional instructors teach the current practice based on the Austrian ski teaching plan. In the ski courses of secondary schools and high schools, this knowledge is passed on to the students by the teaching staff or specially hired ski instructors. Ever since the 1950s, young Austrians have been familiarized with skiing techniques.

Since the 1960s, alpine skiing has become a mass sport for all social classes. This was due to the rapid development of the infrastructure—cable cars, pistes, accommodation, restaurants up to the high alpine zone—which made skiing a pleasure for locals and visitors alike. The sport has changed and evolved, becoming more athletic and sociologically diverse. Snowboarding, for example, is a variant of a youthful lifestyle with its own aesthetics and expression. Fun courses, half-pipes and piste styles offer an alternative to the traditional piste or ski highway, with associations to the early days of skiing, when the aim was to overcome natural obstacles, terrain jumps, etc. at a wild pace. Variant skiing uses the aids to ascent, but then seeks the descent off the piste in open terrain. Whereas touring seeks to minimise the risks of the mountains, freeriding seeks the ultimate challenge, exploring the limits of what is possible. But they all have one thing in common: the mastery of the cultural technique that guarantees the most competent use of a pair of skis and the appropriate equipment in and with winter nature.

A Swing Goes Around the World

Austria and Austrian ski pioneers had a formative influence on the development of skiing in Central Europe. The early history of skiing around the turn of the century into the 1920s was characterized by the search for the optimal ski technique, the best binding and number of skisticks, etc. The two great pioneers of alpine skiing in Austria, Mathias Zdarsky and Colonel Georg Bilgeri, represented different schools of thought and instructed thousands to ski in their respective preferred techniques.

With the first ski films in the mid-1930s, alpine skiing began its spectacular rise, as more and more ski enthusiasts came to the mountains with the first cable cars and chairlifts, and the Arlberg technique became world famous as an alpine form of skiing. The Arlberg Ski Club, the oldest in the Alps, was founded in 1901. Snow became a reference point for modern action, and ski culture in the context of sport, tourism and everyday life became the subject of self-identification for groups and regions. Skiing is not only about skiing in the mountains in winter, but also about combining bodies, knowledge and skills, ways of thinking and speaking into a cultural practice. As early as 1928, the traditional Kandahar Race was held for the first time—organized by Hannes Schneider, exponent of modern alpine ski technique and founder of the first ski school on the Arlberg, and the Briton Arnold Lunn, the creator of modern slalom technique.

With the construction of cable cars and ski lifts in those years, the action within Austria moved increasingly to the western provinces—from Lower Austria to Vorarlberg, Tyrol and Salzburg. At the turn of the twentieth century, the first ski clubs were established, where the technique of alpine skiing was taught—first by Norwegian experts, but increasingly by locals who had studied the textbooks of Zdarsky or Bilgeri (Bilgeri's book *Alpine Skiing* was published in 1910). For a time, there was some dispute as to which was the best technique, but in the end a practice that combined elements of several techniques emerged and was incorporated into the *Austrian Ski Teaching Plan* in the 1930s. The first training course for ski instructors took place in 1927 in Sankt Johann in Salzburg's Pongau region. Zdarsky's Lilienfeld stem turn, its refinement and the Arlberg School's sitting squat-crouch, the development of the basic techniques into the parallel swing and the close ski guidance of Toni Seelos into Stefan

Kruckenhauser's wedel technique with hip bend and heel thrust—these were the elements of the Austrian technique that became the standard for training in other countries as well. Wedeling was replaced by the step change technique in the 1970s, and in the 1990s the carving technique emerged, which allowed for a more intense feeling of acceleration through wider-legged, tighter turning radii. The inventor of this natural form of swinging is considered to be the Viennese ski professor Hans Zehetmayer (1936–2022), who shaped the training of Austrian ski instructors for several decades. For some years now, 'beautiful skiing' has been considered a new trend on the slopes, combining the pleasure and elegance of swings and now represents the measure of all things. Beautiful skiing with close ski guidance was also included in the Austrian ski teaching plan and is taught in ski schools.

Six Austrians Among the First Five

The victories of Austrian ski racers at the Olympic Games, World Championships and World Cup races played a significant role in the rise of skiing. They continue to inspire the younger generation as motivators, their charisma attracting crowds of spectators. Thousands come to the races, millions cheer their idols in front of the television screens, as the races of the Ski World Cup and all major events are broadcast live on Austrian television. Their successes—where a hundredth of a second can mean the difference between victory and hero status, or defeat and ignominy—strengthen national pride and the Austrian soul. Six Austrians among the first five, nine Austrians among the first ten—headlines like these from the late 1990s express dominance, pride and gratitude. Similar in significance to the World and European championship titles of German footballers, the victories of ski stars in the years after the Second World War were pillars of a newly emerging Austrian identity—away from the Danube Monarchy, towards the Alpine Republic. From Toni Sailer to Karl Schranz, Annemarie Moser-Pröll, Franz Klammer, Petra Kronberger, Hermann Maier, Marlies Schild, Marcel Hirscher and Anna Veith-Fenninger—to mention just some of the most famous names—they are Olympic champions, world champions, World Cup winners and role models, and not only for youth. They are also world-renowned ambassadors of the winter sports and tourism country Austria.

Skiing, along with cycling and hiking, is among the most popular sports throughout Austria; and in Tyrol and Salzburg, snowboarding also ranks among the top ten. According to surveys by Statistik Austria and other studies, half of the population identifys as skiers. About a third of the Austrian population ski regularly, with a slightly declining trend in recent years.

Alpine skiing is an indispensable element in winter tourism in Austria. This has a direct value added of around seven billion euros, contributes about 2.5% to the gross domestic product and secures an income for some 185,000 employees. In the western federal states, its economic importance is even greater, as it creates jobs and income also in the peripheral high-altitude side valleys of the Alps.

Therefore, the loss of reputation that Tyrol or Ischgl caused to Austrian tourism during the Corona pandemic came as a hard blow. In this ski resort in the Paznaun valley, also known as 'Luxury Ballermann' some 10,000 ski tourists were infected with the Covid virus, according to the Robert Koch Institute. Après-ski bars and discos reluctantly closed. Almost two weeks earlier, it was already known that an Icelandic travel group had been infected there. However, the state health directorate considered such infection to be "rather unlikely". However, it led to the hasty departure of foreign guests and seasonal workers. The investigative reporting in Germany's largest political magazine *Der Spiegel*—with a cover story about the *Ischgl file* and the failure of the authorities—is likely to have a long-lasting impact, tarnishing the image of Austria as a ski nation.[24]

The question of whether the closure of hotels, discos, and cable cars and the measures taken by the state of Tyrol were implemented much too late will have to be clarified in court. Many of those affected—most of them from Germany—have joined a class action lawsuit for damages, filed by the Austrian Consumer Protection Association. After Tyrol, or rather all of Austria, was temporarily put on the red list by Germany and the borders remained closed, German tourism in Austria came to a near-total standstill. In practice, the cable cars were operating only running for locals in the winter of 2020/21 and revenues dropped by 90%. From the summer of 2021, the race to attract German visitors will be back on track.

There was still a lack of hotels. The most adventurous in the village, the thrill-seekers, those who were not particularly popular, became hoteliers—another step out of the farming society. The inns flourished. The owners began to gain influence in the village. The expanded inn attracts tourists, invites them to come back. They no longer reject the outsider, the stranger. He brings the money, money that has never been so easy to earn.

Characteristic of the years before inflation is therefore the emergence of tourism and its consequences. Summer and then winter sports. This results in a change in leisure time and—as the most important consequence—a transformation of the village society, the formation of a new, enterprising elite of innkeepers or hoteliers. A lively economic activity arises in the

village, profit becomes increasingly important. The connections to the city become increasingly close. … so the farmer seeks the city but above all, the city now invades the village
Lucie Varga, A Valley in Vorarlberg—Between the Day Before Yesterday and Today (1936)

Endnotes

1. An overview of historical and current developments is provided by Kurt Luger & Franz Rest (eds.), *Alpenreisen* (Alpine Travels), Innsbruck: StudienVerlag 2017. On the history of the Alps in particular, see Jon Mathieu, *Geschichte der Alpen* (History of the Alps 1500–1900*),* Vienna: Böhlau 1998; Werner Bätzing, *Die Alpen* (The Alps, Origin and Threat of a European Cultural Landscape*),* Munich: Beck 2015; COTRAO (Working Group of the Western Alps, eds.), *L'homme et les Alpes* (Man and the Alps*),* Grenoble: Glénat 1992.
2. See Werner Bätzing, *Orte guten Lebens* (Places of the Good Life), Zurich: Rotpunkt 2009.
3. See for example Tobias Chilla (ed.), *Leben in den Alpen* (Living in the Alps), Berne: Paul Haupt 2014; Hans Haid, *Vom alten Leben* (Of the Old Life), Vienna: Rosenheimer 1988.
4. For a compehensive overview, see Kurt Luger & Franz Rest (eds.), *Der Alpentourismus* (Alpine Tourism*),* Innsbruck: StudienVerlag 2002.
5. For more details, see Werner Bätzing, Der Stellenwert des Tourismus in den Alpen und seine Bedeutung für eine nachhaltige Entwicklung des Alpenraumes (The Importance of Tourism in the Alps and its Significance for a Sustainable Development of the Alpine Region), in: Luger & Rest (eds.), *Der Alpentourismus* (Alpine Tourism), 2002, 175–196; Orte guten Lebens (Places of Good Living—Visions for an Alpine Tourism between

Wilderness and Amusement Park, in: Luger & Rest (eds.), *Alpenreisen* (Alpine Travels) 2017, 215–236.
6. Kurt Luger, Städter im Kopf? (City Dwellers in the Mind? On the Living Situation of Young People in Pinzgau), in: Herbert Dachs & Roland Floimayr (eds.), *Salzburger Jahrbuch für Politik 1997,* Salzburg: Böhlau 1997, 150–173.
7. Volker Fleischhacker, *Aktuelle Nachfragetrends im Wintersport in Österreich* (Current Demand Trends in Winter Sports Tourism in Austria), *ITR-Tourism Report* 2016, Tulln 2016.
8. See Bätzing in Luger & Rest 2017.
9. See Bätzing, Places of the Good Life, 2009.
10. Richard Schmidjell (ed.), *Wohlfühlwirtschaft, Dienstleister im Wachstumsmarkt Gesundheit* (Feel-good economy, service providers in the growth market health), Vienna: LIT 2008; Alfred Kyrer & Michael A. Populorum (eds.), *Trends und Beschäftigungsfelder im Gesundheits- und Wellness-Tourismus (*Trends and employment fields in health and wellness tourism), Vienna: LIT 2008.
11. http://www.tourismusforschungaustria.at/443273056, 1.9.2021.
12. This is exemplified and meticulously documented in Edith Hessenberger, Walter Hauser & Karl Wiesauer (eds.), *Bau.Kultur.Landschaft im Ötztal* (Building.Culture. Landscape in the Ötztal). Innsbruck: StudienVerlag 2020.
13. Roger Sonderegger & Werner Bätzing, Zweitwohnungen im Alpenraum (Second homes in the Alpine region), *Journal of Alpine Research/Revue de géographie alpine*, 2014; https://doi.org/10.4000/rga.2517. Werner Bätzing, Second homes in the Alpine region, in: *Bergauf,* Issue 3/2020, 70–71.
14. Arthur Kanonier & Arthur Schindelegger, *Raumplanungsrechtliche Ferienwohnungsquote in Vorarlberg* (Spatial planning legal holiday apartment quota in Vorarlberg), TU Vienna (Technical University of Vienna) 2018; Sabine Wüstemann, *Regionale Folgen von*

Landschaftsveränderungen (Regional consequences of landscape changes, A case study using the example of the traditional cultural landscape in the Upper Pinzgau), Land Salzburg: Salzburg 2017.
15. Kurt Luger, Eine Straße auf dem Weg zur Touristenattraktion (A road on the way to becoming a tourist attraction), in: Johannes Hörl & Dietmar Schöndorfer (eds.), *Die Großglockner Hochalpenstraße* (The Grossglockner High Alpine Road, Heritage and Mission*)*, Vienna: Böhlau 2015, 203–230.
16. Dominik Siegrist & Matthias Stremlow (eds.), *Landschaft, Erlebnis, Reisen.* (Landscape, Experience, Travel. Nature-based tourism in parks and UNESCO areas), Zurich: Rotpunkt 2009.
17. *Tiroler Tageszeitung,* 27 July 2016; Disputes and debates about development projects can be followed on various websites, such as www.cipra.org; https://www.alpenverein.at/portal/natur-umwelt/alpine_raumordnung/ski-erschliessungsprojekte/index.php.
18. Moralisch fragwürdiger Skitourismus in Zeiten des Klimawandels (Morally questionable ski tourism in times of climate change), in: *LUA-Notes* 4/2020—Landesumweltanwaltschaft (State Environmental Advocacy Salzburg) www.lua-sbg.at; see also Thema aktuell: Ist der Kunstschnee ein Umweltproblem? (Is artificial snow an environmental problem?), in: *Salzburger Nachrichten,* 4.1.2021.
19. Data on this can be found on the website of the Austrian Chamber of Commerce, www.wko.at.
20. For a detailed discussion, see Kurt Luger & Franz Rest, Alpenreisen-Alpentourismus (Alpine Travels—Alpine Tourism, Positioning with retrospective and long-term view), in: Luger & Rest, *Alpenreisen* 2017, 15–40.
21. OcCC/ProClim—Advisory body for questions of climate change/Forum for Climate and Global Change (ed.), *Klimaänderung und die Schweiz 2050* (Climate change and Switzerland 2050, Expected impacts on

environment, society and economy), Berne 2007; MeteoSwitzerland, *Technical Report* 254, Mild and cold mountain winters; www.meteoschweiz.ch, accessed 1.3.2017.
22. Robert Steiger & Bruno Abegg, Klimawandel und Konkurrenzfähigkeit der Skigebiete in den Ostalpen (Climate change and competitiveness of ski areas in the Eastern Alps), in: Egger/Luger (eds.), *Tourismus und mobile Freizeit* (Tourism and mobile leisure), Norderstedt: BoD 2015, 319–332; Ulrike Pröbstl-Haider/Dagmar Lund-Durlacher/Marc Olefs/Franz Prettenthaler (eds.) *Tourismus und Klimawandel* (Tourism and Climate Change). Open Access 2021. https://doi.org/https://doi.org/10.1007/10.1007/978-3-662-61,522-5; Der Österreichische Tourismus im Klimawandel (Austrian tourism in climate change), https://ccca.ac.at/wissenstransfer/apcc/broschuere-der-oesterreichische-tourismus-im-klimawandel, 4.8.2021.
23. The following presentation is largely based on the following sources: Heinz Polednik, *Weltwunder Skisport* (World Wonder Skiing), Wels: Welsermühl 1969; Joachim Glaser, *Goldschmiede im Schnee. 100 Jahre Salzburger Landes-Skiverband* (Goldsmiths in the Snow. 100 Years Salzburg Province Ski Association 1911–2011), Vienna: Böhlau 2011; Sabine Dettling/Bernhard Tschofen, *Spuren, Skikultur am Arlberg* (Traces, Ski Culture at Arlberg), Bregenz: Bertolini 2014; Elisabeth Längle, *Der Arlberg—Natur- und Kulturlandschaft* (The Arlberg—Nature and Cultural Landscape), Brandstätter: Vienna 2011; Reinhard Bachleitner (ed.), *Alpiner Wintersport* (Alpine Winter Sports), Innsbruck: StudienVerlag 1998.
24. Der Spiegel no. 27, June 27, 2020.

8

Tourism as a Development Perspective

A traveller in Africa
Once saw no lions wide and far.
But eager to glorify his ownself,
he wants to report at home,
that he could have—what everyone grants him –
easily been eaten by real lions.
Eugen Roth, *Beinahe* (Almost), from the German original

"It's quite a remote area here!" Thundu Sherpa, village teacher and owner of a small lodge in Simigaon (lit.: bean-village), at the foot of the sacred mountain Gauri Shankar, is happy to see tourists. A few hundred come up each year to visit the quiet Rolwaling Valley, on the border between Nepal and Tibet. A small additional income is generated for the family, but not much more, as most tourists book a fully organized trek in Kathmandu or are on their way to one of the eight-thousanders beyond the Trashi Laptsa Pass in Sagarmatha (Mount Everest) National Park. Everything, even the fried eggs, is portered in from the capital. Thundu also rents out one of the few flat places where tourist tents can be pitched. Right next to it are the millet fields of his father, who once was a mail-runner for Edmund Hillary, the first summiteer of Mount Everest, in 1953.

The Rolwaling Eco Tourism Project (RETP), implemented from 1996 to 2008 by the Austrian-Nepalese NGO EcoHimal, has created the structure and basic infrastructure for gentle ecotourism in this region. The original plan was for a two-week trekking route on existing paths through villages and the terraced landscapes that show the beauty of the region but also the hard work of cultivating these mountain lands. However, the project was hampered by a ten-year civil war and later by the construction of a large hydropower plant nearby. This also changed the interests of the residents and many were recruited as construction workers. Due to the preliminary work done by the project, however, the Gauri Shankar Conservation Area was created in 2010, increasing the proportion of protected areas in Nepal to almost a quarter of the total land area.

A road blasted into the landscape now leads through the thick vegetation to the power plant, bringing many city dwellers to visit the area. While trekkers used to come mainly from the Western industrialized countries, now the Asian middle classes are attracted, and young Nepalese are discovering their homeland. The valley of the Tamba Khosi River, which originates in Tibet, and the Rolwaling Valley have become popular for short vacations. But the great earthquake of 2015, with its epicentre in this area, was catastrophic. A new lodge in Simigaon fell victim: the quake measured 7.8 on the Richter scale: a major earthquake. Across Nepal, tens of thousands of people were killed and hundreds of thousands of families lost their homes, livestock and livelihoods.

Simigaon and the entire Dolakha district, some 180 km northeast of Kathmandu, have still not fully recovered from this blow. The subsistence farming of the farmers had been precarious, soil fertility been declining for years, and with harvests often inadequate to feed large families for a whole year. The tourism promoted in the RETP was originally intended as additional income for the farming families, to help reduce out-migration pressures. However, visitor numbers have dropped, and many men are still forced to earn money elsewhere. They work in tourism, as porters, cooks, trekking guides, and as cheap labour for building football stadiums and skyscrapers in countries on the Arabian Gulf; they toil as harvest helpers on plantations, or as day labourers. Not infrequently, they never return to their wives and children, disappearing somewhere in search of a better life.[1]

A better life—that's what the mountain farming families in the Himalayas dream of, like millions in poverty-stricken regions around the globe. Whether in the lush tropical jungle, on white sandy beaches or in the shadow of the mountain giants—the landscape is breathtaking, ideal for holidays, adventure, relaxation. But what do the often desperately poor people living there get from this beauty? It is the investors and tour operators in the metropolises, such as the trekking agencies in distant Kathmandu and abroad, who benefit most. Tourism has also created jobs locally and is economically significant: but can tourism alleviate poverty? How should such tourism be crafted for the benefits to reach the local poor population?

Locals and Foreigners on the Path to Sustainability

Until recently, the general assumption in the development debate was that tourism in developing societies causes more harm than good, or is too risky an economic sector to contribute to long-term livelihood security or improved living conditions. This is certainly true in many places, but it is an untenable blanket condemnation. Ultimately, it depends on how tourism is practised: the pro-poor tourism model, aims to benefit the poor and redistribute profits to local people to improve their living conditions. Together with several other NGOs and the UN World Tourism Organization WTO, an action plan titled ST-EP—sustainable tourism, eliminating poverty—was developed, implemented during the 2000s. Many major development organizations, including such as the World Bank, the Asian Development Bank, UNDP, the British DFID/ODI and the Dutch volunteer organization SNV participated, or started their own projects, having recognized that carefully designed tourism can be well integrated into regional development projects, with more positive than negative overall effects. The basic trend in development cooperation—to focus more on job creation and business start-ups (small and medium-sized enterprises), to stimulate local economies and promote private-public partnerships—has contributed to this re-orientation.[2]

Today, we can conclude that poverty-reducing tourism benefits the local populations in developing countries if they are fully involved and a redistribution mode is applied. *Pro-poor tourism* is therefore not a specific tourist product or a sector niche, and it should not be equated with soft tourism or eco-tourism, although many projects practice such forms. Rather, it involves a concept of tourism development and management, a strategic orientation that puts the development policy dimension at the centre. It promotes the links between the tourism industry and the local people, who are given more opportunities to participate in product development and as service providers. There are many possible forms of application and strategies, ranging from job creation to training measures to participation models. Almost any company can be integrated into this strategy—a small lodge, a city hotel, a tour operator, even a company that builds the infrastructure. The critical factor is not the type of company or the form of tourism, but the visible increase in net benefit for the poor population in the region where tourism is practised.[3]

The basis of the UN-WTO's commitment in the sector of development cooperation through poverty-reducing tourism was a pragmatically-oriented study published in 2002, titled *Tourism and Poverty Alleviation*. It views tourism as a top export product for developing countries and Least Developed Countries (LDCs): tourism shows good growth rates and has developed into an important source of foreign exchange income in many countries. Some 80% of the world's poor (defined as those who had to subsist on less than one US dollar a day) lived in 12 countries: and in eleven of them, tourism played a significant and growing role. The UN-WTO was attracted by the idea of turning the poor into creators and exporters of a well-considered product, and also with the development potential that can draw on a vast diversity of ethnic groups, biodiversity and landscapes.[4]

To unleash this potential, village cooperatives and small businesses were given access to the tourism market. To reduce large leakage effects, they were networked with the existing industry. Complementary to the subsistence economy in developing societies, such projects promote small and medium-sized enterprises and ensure that the income from tourism actually flows back into the region where it was earned. However, this inevitably leads the discussion to where other development efforts often end: in the corruption swamp of national elites and governments, in the unequal power structures, in the completely unequal access to education and development, in the highly unjust distribution of property, land and infrastructure.

Such ambitious tourism projects work best in specially created conditions, with continuous monitoring and clear development policy structure or integration of tourism policy measures into long-term regional development programmes. Development organizations assume that tourism, like road or power-plant construction, education and many other factors, plays an important role in changing living conditions in poor countries. Sustainability strategies must therefore be regionally anchored and responsive to local conditions. Where there is potential for tourism expansion, there is nothing to prevent this economic sector from being promoted.[5]

Today, three decades after the UN Conference on Environment and Development in Rio de Janeiro 1992 (Earth Summit, Agenda 21) and after numerous consultations of the UN Commission on Sustainable Development (CSD 7), the topic of tourism and sustainable development has found expression in the Sustainable Development Goals (SDGs) and in UN declarations; it is included in many demand concepts and programmatic documents such as Agenda 2030. The ecological aspects as well as the participation of local stakeholders are included in tourism sustainability concepts, as is the major goal of *poverty reduction*.

In defining objectives and steps towards achievability, UN-WTO and development organizations argue very similarly: Pro-poor tourism enables the protection of natural, historical, cultural, and other resources for the future, but yields immediate benefits as well. Such development must be carefully planned and well managed so as not to cause environmental damage or socio-cultural problems. The high quality of the tourist product must be developed, achieved and permanently secured, because this is how the destination achieves or retains its attractiveness. The benefits must reach as many people in society as possible. This also addresses Fair Trade considerations, which have been disseminated in the German-speaking world and beyond by the Swiss working group for Tourism and Development in Basle under the slogan *Fair Travel*.[6]

The importance of sustainable tourism as a development policy measure is underscored by the resolution adopted by the UN General Assembly in December 2020. Under the heading *Promotion of sustainable tourism, including ecotourism, for poverty eradication and environment protection*, the *2030 Agenda for Sustainable Development* is highlighted as a guideline for tourism, while it is also emphasized that sustainably oriented tourism contributes to the achievement of the UN Sustainable Development Goals. This kind of tourism can help to combat poverty, improve the livelihoods of the local population, and contribute to the preservation of biodiversity. This UN Resolution was preceded by a global evaluation process in which the progress of tourism projects in the context of climate change and Agenda 2030 processes was presented. Even if the creation of concepts or management plans was already described as major progress in some places (intentions recorded on paper), we can note a gradual rethinking in global tourism, as numerous projects show.[7]

The Measurable Benefit: the Case of SNV

John Hummel has led Pro-poor tourism projects for the Dutch development organization SNV in various Asian countries for over 15 years. He sees them as a successful economic and development policy instrument in the fight against poverty. Integrated into regional development projects, referring to his assignments in Nepal, Bhutan, Vietnam, and Laos. SNV has consistently pursued the path of supporting job creation and business start-ups (small and medium-sized enterprises) in development cooperation, stimulating local economies, and promoting private-public partnerships. Due to the differing sizes of the projects and the number of beneficiaries, it is not possible to answer the question of effectiveness on a purely statistical basis. However, prosperity indicators give a clear indication that income triggered by tourism brings more than just monetary benefits. Villages with tourism generally have more solidly built and better-equipped houses and more livestock; their children attend school regularly, and the local people rate their supply situation as significantly less problematic.[8]

The range of approaches also varies greatly from country to country. SNV worked in Nepal on the development of trekking routes, was involved in small-scale training, in consulting on destination development, in building a multi-stakeholder structure—always focused on poverty reduction. In Laos, the organization was involved in the development of a national tourism plan and in marketing the country as an ecotourism and trekking destination. At the local level, the tourism organizations were advised and employees trained; also value chains were developed, from which the local population derived income—as providers of accommodation and meals, as producers and sellers of handicrafts, and as tour guides or providers of excursions.

In Bhutan too, the concept was pursued to develop two trekking routes and to involve the local population in the villages accordingly. Households directly involved were able to increase their incomes significantly, but it was not guaranteed that the truly poorest could also benefit. Such doubts could not be completely dispelled even by later project evaluations. After budget cuts, SNV withdrew from the area and focused on agriculture, sanitation and hygiene, or health projects, seeking a more direct connection with poverty reduction.

In fact, the SNV's multi-year country programme for Vietnam went far beyond small-scale tourism development. The strategic plan also included elements of agrotourism; it supported value chains, and promoted pro-poor sustainable tourism at the national and village levels. From these approaches, a comprehensive tourism concept developed in the wider Mekong Sub-Region with the support of the Asian Development Bank, in which the development policy goal of poverty reduction plays a central role.[9]

Nepal: the Bird of Paradise in Himalayan Tourism

Eight of the 14 Himalayan eight-thousanders are located within the territory of the former Hindu kingdom, today's Republic of Nepal. The country, about which little else was known, made early headlines in the world media due to impressive first ascents. In 1951, the first man—a team of French and local people—set foot on Annapurna I. In 1953, the first successful ascent of Mount Everest/Sagarmatha took place, and in 1954, Asian traveller

Herbert Tichy and a small Austrian-Nepalese team stood on Cho Oyu, at 8188 m the sixth highest mountain in the world. The pioneering phase of high-altitude mountaineering was concluded in 1960 with the first ascent of Mt Dhaulagiri. The Swiss expedition team included an Austrian, Kurt Diemberger, who first climbed two eight-thousanders and is now over 90 years old.

Expedition tourism provides the state and many locals with a permanent source of income to this day. However, numerically and economically significant mountain tourism did not begin until the 1970s. Prior to that, the exotic charm, tiger hunts, and the sacred architecture in the Kathmandu Valley attracted an exclusive international audience. For a short time, Nepal was a top destination for the 'flower children', because hashish (*ganja*) thrives there. However, the authorities put an end to this low-budget tourism by imposing heavy penalties for drug use and trafficking. In addition to high-altitude mountaineering, trekking tourism developed dynamically. Mountain hiking over high passes and from village to village in the shadow of the eight-thousanders brought more and more visitors and foreign currency into the country, year after year.

As early as 1973, Nepal's first national park, the 932 km^2 Royal Chitwan National Park, was established in the jungles of the southern lowlands. Wildlife resorts within the park boundaries became ideal starting points for jungle safaris, a major tourist attraction, although their ecological compatibility remains controversial to this day. In 1976, the intervention of international experts led to the establishment of the Sagarmatha (Mount Everest) National Park. The fragile alpine ecology was to be protected from the excesses of tourism as well as from the locals, who depend on resources from the forest for their subsistence economy. The high-alpine landscape with its short vegetation period cannot withstand this double burden in the long run. In 1979, the national park was declared a UNESCO World Heritage Site, an award that has come under criticism recently because the interventions in nature have reached almost unacceptable levels. In addition, the region is heavily affected by climate change; melting glaciers and the outbreak of glacial lakes threaten the rich biodiversity and the habitat of humans and animals.[10]

The establishment of the Annapurna Conservation Area Project in 1986 was a milestone. This popular trekking area (as popular as the Everest region) was transformed into a protected zone and its residents were trained in resource conservation. At the same time, trekking tourism was promoted and a tourist product was developed that took local circumstances into account. The mountain farmers and the state were able to benefit: through income from tourism (direct income, trekking fees, entry into the protected zone), through the preservation of their environment, through an improved understanding of ecological interrelationships, through the advice and accompanying control of the project team, which promoted well considered regional development. This project was under the aegis of the King Mahendra Trust for Nature Conservation (today: National Trust for Nature Conservation), a major national environmental organization, in which the last king of Nepal served as chairman for several years.[11]

Most trekking tourism in Nepal today takes place in national parks and designated protected areas. In these impressively scenic areas, local subsistence farmers earn additional income from tourism. In Khumbu, the region around Mount Everest, and to some extent in the Annapurna area, tourism has gradually become dangerously dependent on international developments. Agriculture is difficult in high-elevation villages, so tourism became the main source of income for mountain farmers living above 3000m.

Business flourished until the year 2000, with 490,000 international tourist arrivals, and some 100,000 tourists visiting the mountains. Then, however, tourism plummeted and political events in Nepal made the headlines. Asian tourism suffered massively from pandemic events such as bird flu and SARS, 9/11 and the Afghanistan war, accompanied by fears of terror and religious fundamentalism. Nepal also had its own problems. Foreign visitors were deterred by the ten-year civil war (1996–2006), long periods without democratically legitimized governments, as well as curfews and forced strikes that repeatedly paralysed the country. In 2002, only 275,000 tourists arrived. Only after the 2006 peace agreement between the royal house, the parties represented in the parliament, and the Maoist rebels, and the founding of the republic (Nepal had been a monarchy until then) did tourism pick up. However, there ensued a series of earthquakes in 2015, with major damage throughout the country. Before the Corona pandemic, which brought almost a complete standstill, approximately one million tourists arrived: about one-sixth of them set off for mountain trekking tours.[12]

From the experiences of more than three decades with village tourism and trekking, not only the locals have drawn their lessons, but also those responsible in the national tourism authority and in the ministries involved. Tourism has been given high priority in national development plans, recognising that it can benefit both the local and national economy. In a country with few natural resources other than wood and water, there are not many alternatives. However, sustainable tourism had gradually become established and was recognised as a tool for sustainable development by international donor institutions such as the Asian Development Bank and international development organisations. In 2004, Nepal designed a Pro-Poor Sustainable Tourism Policy, a Nepal Tourism Industry Strategic Plan and a marketing strategy that was to shape the country's tourism appeal until 2020. For the first time, tourism was included in the district-level development plans in order to involve the regional authorities in the tourism development process. All of this shows that the development of sustainable tourism is now accorded high importance.

Development at the Level of Village Communities

The Rolwaling Eco Tourism Project (RETP)[13] was also committed to these goals, although the pro-poor approach was initially not yet part of the development policy canon. It started with a feasibility study in which the local population expressed their wish for the development of an infrastructure for village tourism. The RETP, like all other EcoHimal projects, was to be distinctly participatory (motto: *Where there is unity, there is energy!*), contributing to environmental protection and reaching as wide a circle of beneficiaries as possible. A self-administration body was created as the carrier organization for the 24 villages involved, whose members had received special training. Nevertheless, the final evaluation showed weaknesses in self-administration. In summary, we can say that the RETP was a rural livelihood project, an integrated regional development project that envisaged the establishment of trekking tourism with basic infrastructure as a specific component of livelihood.

However, more than 15 years after the end of the project, it is clear that the model of lodges owned by the village community has not proven itself. The civil war and the major construction site in the region were probably the main obstacles to this. In addition, tourists stayed away in this first phase, which discouraged the local communities. But those involved in the project have learned how tourism is practised, how to handle waste and wastewater responsibly, the benefits of compost toilets, and how sanitary facilities lead to appropriate hygiene behaviour. At the beginning of the project, there were hardly any functioning toilets, but by the end, over a thousand septic tanks had been built: indeed, the demand was so great that only subsidies for building materials were granted. Initially, more than two-thirds of the population could neither read, write nor do basic arithmetic; the project offered a broad literacy programme, supplemented by special training courses in which tourism-relevant knowledge and skills were imparted.

The positive effects on the quality of life in the region are unmistakable, even without the originally targeted significantly higher tourism volume. The supply of clean drinking water to the villages—altogether 68 km of water pipes were laid—has significantly eased the daily life of women, who used to spend hours fetching water. The connection between hygiene, water quality, nutrition, and health was the subject of many training courses, in which large parts of the population participated. These courses were conducted in connection with the toilet program, the establishment of kitchen gardens, and health camps held in the villages. In these remote regions of the Himalayas, where there is hardly any medical care, health prevention is a top priority. The RETP found recognition and followership among the broader population, as it became clear that not everyone would benefit from tourism as such.

The good reputation was also due to the fact that the project staff did not leave the region, which was repeatedly marked by fighting, also during the civil war. However, this trust also led to the general view that "the project people will take care of it"—and work and responsibility were gladly left to the project administration. This also had to do with the tense political situation. Local cadre members and sympathizers of the Maoist rebels were represented in the village communities from the beginning, and soldiers of the rebel army patrolled some meetings, their Kalashnikovs slung over their shoulders. At the end of the civil war, however, the EcoHimal project management was able to hand over responsibility for the project to the umbrella organization of the village communities, and with the easing political situation, a growing understanding of ownership developed among its members.

Through the project and its impact on the development of the region, ecotourism gained importance also at the district level. The administration proposed to define the entire region as a protected zone, as a Conservation Area. This was taken up and the government, largely for publicity reasons, declared the implementation in its cabinet meeting prior to the Copenhagen Climate Summit in December 2009, which it held at Mount Everest Base Camp. Various concepts and strategy papers developed in the project served as aids for planning and the tourism master plan of the region. Just as Nepal today holds a leading position in ecotourism in Asia, so are Nepal's northern Dolakha and Sindhupalchowk districts among those regions where efforts towards sustainability in tourism are clearly visible.[14]

Integrative Approach: linking Tourism and Agriculture in the Caucasus

Development policy organizations have practiced various promising approaches, discarding them if they did not yield the expected results. Tourism aimed at poverty reduction or a sustainability perspective has gained recognition as a suitable instrument in the development of rural or remote areas. This approach being applied in a recently completed project to strengthen mountain tourism in the Caucasian republic of Georgia. In a multi-year project co-financed by the European Union, Austrian Development Cooperation with German, Swedish, and local partners is attempting to improve tourism management, services, product quality, and marketing in two mountain regions. A sustainable tourism strategy for these areas aims to connect economic, social, cultural, and ecological aspects as much as possible in order to improve living conditions locally in the long term.[15]

Sustainable tourism and *organic farming* are at the heart of this pilot project. When related and coordinated, they have positive effects on local biodiversity and increase resilience to climate change. Organic farming is an efficient way to integrate small farmers into value chains on local and international markets, diversify their income opportunities, reduce unemployment and poverty, and promote economic growth.

The mountain regions of the Caucasus, with their impressive nature and traditional village life, have great tourist potential, particularly for rural eco- and agrotourism. Nature and adventure, skiing and mountains, cultural heritage and wine are the central tourist offers in the pilot region, which includes parts of Upper and Lower Svaneti and Imereti. Svaneti is the favorite region for mountaineers and hikers. The popularity of this region and especially the village of Ushguli has significantly increased because it was advertised by international tour operators as the highest permanently inhabited settlement in Europe (Georgia is considered 'the balcony of Europe'). Since 1996, with its distinctive defense towers Upper Svaneti has been a UNESCO World Heritage Site. Tourist products (organized hikes, trekking, ski tours with mountain guides, luggage transport with horses, and overnight stays in traditional huts and farms) offer numerous job opportunities for the local population.

The pilot project, which is being implemented with reference to national master plans and EU programmes, focuses on the cooperation of various actors and the integration of activities in a cluster model for sustainable development of the destination.

A *cluster* is understood as a group of interconnected companies in the same region (suppliers, producers, trade, etc.) under one roof. The management coordinates their cooperation with national and local authorities, educational institutions, and also partners from other sectors. Their activities are regulated by a common action programme. Recently, this approach has also been applied to the tourism sector. Thus, a *tourism cluster* includes a group of tourism-related companies with the aim of supporting development through the use of specific management techniques. A tourism cluster also aims to increase the competitiveness of the participating partners through cooperation and interaction by enabling shared learning and knowledge exchange.

In the pilot region, a Destination Management Structure (DMO) has been established to coordinate this integration of various actors, promotes partnership, and destination management. There are no experiences with local initiatives like in other regions of the Caucasus, which support the mechanisms of the participatory approach in local decision-making processes and the sustainable growth of the districts, in the pilot region. Initiatives must also be created that strive to mobilize the community—with a particular focus on the inclusion of young people, women, and people from peripheral areas. This requires a common vision and a long-term strategy for integrative development in the mountain regions.

The pilot project aims at sustainable tourism development. This also requires the collection of valid data in order to obtain well-founded evidence on the footprint of each activity or the impact of each innovation. Such data must provide information on tourism flows, the performance of businesses, the quantity and quality of employment and social aspects, the tourism supply chain, etc. Sustainable development requires evaluation and monitoring of waste management, energy consumption, landscape protection and biodiversity, to name but a few. In the long term, monitoring could be carried out through the introduction of an indicator system such as the European Indicator System (ETIS).[16] This requires reliable and comprehensive statistical data, which is currently lacking in many tourism regions.

Showcase Models of Good Practice

The number of so-called Pro-poor-projects—even if they are not explicitly identified as such or do not operate under the patronage of the UN-WTO as sustainable tourism, eliminating poverty (ST-EP) projects—has increased enormously worldwide over the past 25 years. This is related to the introduction of the Millennium Development Goals (MDGs) or the Sustainable Development Goals of the UN (SDGs), which are the measure of all things in the orientation of regional development efforts and to which the signatory countries have committed themselves. They are therefore of utmost importance for tourism. Particularly addressed are goal 1: reduction of poverty, 2: ending hunger and developing sustainable agriculture, 10: reducing inequalities, 12: shaping production and consumption sustainably, 13: combating climate change, 15; preserving biodiversity, and 17:

developing international partnerships, in order to achieve all goals, because tourism permeates all areas of life and is closely linked to societal development.[17]

One example is the *To Do!* competition organised by the German Studienkreis Tourismus und Entwicklung (Starnberg Circle for Tourism and Development). Every year since 1995, at the ITB, the largest tourism trade fair held in Berlin, a few outstanding, sustainability-oriented tourism projects have been honoured as examples of good practice and evaluated by a team of experts. These projects do not all have a large economic impact, as they often cover a very small area, but they all demonstrate an effort to meet the criteria of sustainable development.[18]

In 2021, the community-based tourism project *Rutas Ancestrales Araucarias* in Chile received this award. On seven routes, visitors are introduced to the culture and way of life of an ethnic group, the *Mapuche*, the *people of the earth*; 30 partner organizations look after guests who live with families, eat with them and visit their farms, learn about local healing and medicinal plants as well as the area on the edge of the Villarrica National Park. Altogether 38 families can generate direct income from this project, but even more important is the project's contribution to strengthening their own identity, as the population has been fighting for recognition of their rights for centuries and is still oppressed by the state, their habitat reduced by military interventions.[19]

In 2020, the *Esfahk Historic Village* in Iran and the *Banteay Chhmar Community Based Tourism Project* in Cambodia received this award. The twelfth century Banteay Chhar temple complex is one of the country's most important cultural treasures. Here, 90 families of the village give visitors insights into village life through homestays. Supported by the French NGO *Agir pour le Cambodge,* a village development fund was established, which finances a waste disposal program and training measures. Tourism helps to preserve the temple complex and provides additional income for the families involved.

In 1978, a severe earthquake destroyed many traditional mud villages in the eastern Iran, including the desert village of Esfahk. The survivors built a new village, just a few hundred meters from the old one. In 2009, a village committee was formed, which reconstructed the old village and, with the help of local architects and experts from Tehran, created a touristically usable and cultural-historical gem. The guest houses have been built in the old mud style but are earthquake-proof; there is the traditional mosque, the hamam, a restaurant that offers local dishes, a souvenir shop that sells typical and locally produced handicraft items. Over the years, a sustainability-oriented employment model for 50 people has emerged from a project based on volunteerism. These operate the guest houses and convey the 'old life' in this desert district located away from major pilgrimage and tourism regions. A quarter of the 800 villagers benefit directly or indirectly from the community project. Many young women are involved, selling handicraft products, herbal mixtures and running the restaurants. Tradition lives on, also in terms of building materials or the irrigation of fields, which receive just enough water through small canals to produce food without artificial fertilizers.

Similarly remote is the Uzbek village of *Mitan,* where the travel agency Silk Road Destinations developed a community-based project that was awarded in 2014. Here, special emphasis is placed on the involvement of young people. The small village in the steppe offers visitors direct contact with the population and exemplifies criteria such as participatory approach, environmental compatibility and economic usefulness.

The *Maquipucuna Ecotourism Project* in Ecuador promotes the protection of biodiversity through local development and ecotourism. A 6000-hectare nature reserve was designated, an eco-lodge with ecotourism activities was set up and young people received training in environmental topics. The project received the TO DO Award in 2018.

In the same year, the *!Khwa ttu San Culture and Education Centre* in South Africa was also recognized. The aim of this organization is the restoration and preservation of the *San* ('Bushman') culture, whose habitats extend across several countries of southern Africa. The San Culture and Education Centre near Yzerfontein serves as a restaurant with various accommodation options as a tourist provider. It is also a vocational training center for young San members, to provide them with professional future perspectives after phases of uprooting and displacement, so that they can work in their home regions in tourist professions. As a third focus, !Khwa ttu Centre is committed to sustainable nature projects, and promotes cultural measures for the preservation and learning of the San culture.[20] Ethnotourism in the Kalahari and cultural tourism projects with the San, Khomani or Bushmen of South Africa have received massive media support in recent years. The stereotype of the 'mystical people' undermines the idea of a young and educated generation, which requires a critical view of this development.[21]

The Starnberg Circle for Tourism and Development awards tourism projects that closely involve the local population in their planning and implementation, use resources in an environmentally friendly way and create sustainable economic cycles. This creates new perspectives, strengthens self-efficacy and cultural identity, promotes equal opportunities and social progress. This also includes fair working conditions, educational opportunities and social security. Over the years, more than 500 projects have been submitted and 54 awarded, most of them from developing or emerging countries. As diverse as the concepts are, they have one thing in common: They help locals to use tourism for their development. And the travellers also benefit: They get to know countries and people in an authentic way.

Travel, reading and reflection—these are the three sources, from which I draw when writing, they provide me with the material. In addition, sporadic excursions into poetry and photography help me.

The first source is therefore the journey as discovery, as exploration, as the effort of the scientist.

Travel as a search for truth, not for relaxation.

I want to approach the reality that I encounter. To see it, recognize it, understand it. This requires constant concentration and at the same time a constant opening up, in order to absorb, experience, remember as much as possible.

Ryszard Kapuscinski, Die Welt im Notizbuch (The World in a Notebook)

Endnotes

1. On the harsh living conditions in the Himalayas see Kurt Luger, *Auf der Suche nach dem Ort des ewigen Glücks* (In Search of the Place of Eternal Happiness: Culture, Tourism and Development in the Himalayas), Kathmandu: Vajra 2014.
2. A detailed discussion of this topic can be found in Kurt Luger, Ökotourismus, Partizipation und nachhaltige Entwicklung (Ecotourism, Participation and Sustainable Development—Experiences from a Regional Development Project in Nepal), in: *TW-Zeitschrift für*

Tourismuswissenschaft (Journal for Tourism Science), 2/2010, 165–183; see also Melanie Ströbel, Tourismus Forschung (Tourism Research), in: Ursula Kluwick & Evi Zemanek (Eds.), *Nachhaltigkeit interdisziplinär* (Sustainability Interdisciplinary, Concept, Discourses, Practices), Vienna: UTB 2019, 242–261.
3. For an overview, see Regina Scheyvens, Exploring the tourism-poverty Nexus, *Current Issues in Tourism*, 2–3, 2007, 231–254, http://dx.doi.org/https://doi.org/10.2167/cit318.0.
4. Since then, the number of such studies has grown enormously, with a wide range of case studies that provide empirical evidence that this approach can lead to the desired successes. The UN WTO has published a manual for decision-makers and tourism developers that outlines good practice in the approach: *Manual on Tourism and Poverty Alleviation—Practical Steps for Destinations*, 2010; https://www.e-unwto.org/doi/book/https://doi.org/10.18111/9789284413430, accessed 8.8.2021.
5. John Hummel, Pro-poor Sustainable Tourism, in: Kurt Luger, Christian Baumgartner & Karlheinz Wöhler (Eds.), *Ferntourismus—wohin?* (Long-distance tourism—where to? Global tourism conquers the horizon), Innsbruck: StudienVerlag 2004, 123–146.
6. Several NGOs have been advocating for fair travel for years and have established an international network. See https://www.fairunterwegs.org/; https://www.tourism-watch.de/de; https://www.nf-int.org/
7. https://digitallibrary.un.org/record/3879234; https://undocs.org/en/A/RES/75/229
8. John Hummel, Tara Gujadhur & Nanda Ritsam, Evolution of Tourism—Approaches for Poverty Reduction Impact in SNV Asia. Cases from Lao PDR, Bhutan and Vietnam, in: *Asia Pacific Journal of Tourism Research* 4/2012, 369–384; https://doi.org/https://doi.org/10.1080/10941665.2012.658417.

9. John Hummel & Rene van der Duim, Tourism and development at work: 15 years of tourism and poverty reduction within the SNV Netherlands Development Organisation, in: *Journal of Sustainable Tourism*, 3/2012, 319–338. https://doi.org/https://doi.org/10.1080/09669582.2012.663381.
10. See Ang Rita Sherpa, *Climate Change in the Himalayas*, The Partners Nepal: Lalitpur 2020; Jack Ives, *Himalayan Perceptions, Environmental Change and the Well-being of Mountain Peoples*, London: Routledge 2014; Sanjay Nepal, *Tourism and Environment, Perspectives for the Nepal Himalaya*, Kathmandu-Innsbruck: Himal Books/StudienVerlag 2003; Sushma Bhatta, Robin Boustead & Kurt Luger, The Highest Mountain in the Shadow of Climate Change: Managing Tourism and Conservation in a World Heritage Site, Sagarmatha National Park, Nepal, in: Marie-Theres Albert, Roland Bernecker, Claire Cave, Anca Claudia Prodan & Matthias Ripp (eds.), *50 Years World Heritage Convention: Shared Responsibility—Conflict & Reconciliation*, Springer Heritage Studies: 2022. https://doi.org/https://doi.org/10.1007/978-3-031-05,660-4.
11. Siddharta Bajracharya, The Annapurna Conservation Area Project, in: Patricia East, Kurt Luger & Karin Inmann (eds.), *Sustainability in Mountain Tourism: Perspectives for the Himalayan Countries*, New Delhi: Pilgrims 1998, 243–254.
12. Data from the Ministry of Tourism and Civil Aviation; further data: https://de.statista.com/statistik/daten/studie/425075/umfrage/anzahl-der-internationalen-touristenankuenfte-in-nepal/ 8.8.2021.
13. Detailed information about the project can be found on the EcoHimal website, http://ecohimal.org/index.php?id=20&L=128.
14. See Luger, note 2; see also Birgit Brosio, *Entwicklung von Indikatoren zur Messung von nachhaltigem Tourismus in Bergregionen am Beispiel der Gauri Shankar-Region*

in Nepal (Development of Indicators for Measuring Sustainable Tourism in Mountain Regions, A case study of using the example of the Gauri Shankar Region in Nepal). PhD dissertation, University of Salzburg 2016.

15. For information about the GRETA project of Austrian development cooperation, in which I was involved in the concept development, see https://eu4georgia.ge/green-economy-sustainable-mountain-tourism-and-organic-agriculture-greta/
16. https://ec.europa.eu/growth/sectors/tourism/offer/sustainable/indicators_en
17. https://www.unwto.org/sustainable-development
18. www.to-do-contest.org
19. https://www.todo-contest.org/pramierte-projekte/rutas-ancestrales-araucarias-chile/
20. https://www.todo-contest.org/pramierte-projekte/khwa-ttu-san-culture-and-education-centre/;
21. See Nhamo Mhiripiri & Keyan Tomaselli, Language Ambiguities, Cultural Tourism and the Khomani, in: Kurt Luger & Karlheinz Wöhler (eds.), *Kulturelles Erbe und Tourismus* (Cultural Heritage and Tourism), Innsbruck: StudienVerlag 2010, 285–295.

9

Paths to Sustainability

*It is essential to reshape the tourism sector in a safe,
fair and climate-friendly manner.*
Antonio Guterres, UN Secretary-General

*We are determined to take the bold and transformative steps that are urgently needed,
to shift the world onto the path of sustainability and resilience.*
United Nations: Transforming our World. The 2030 Agenda for Sustainable Development.

*I hope my microphone was on.
I hope you all could hear me.*
Greta Thunberg, Houses of Parliament, London, 23 April 2019

Development with substance and reflection, with good, lasting solutions for the positive development of the planet and its inhabitants: this, no more and no less, is the task facing us all. With the *Agenda 2030*, the international community has set 17 goals for socially, economically and ecologically sustainable development. They apply universally and equally to all countries, and range from the elimination of global hunger to the strengthening of sustainable consumption and production to measures for climate protection. The signatory states have agreed to this politically ambitious agenda.

The 2021 *Dasgupta Review* on the economics of biodiversity and the massive misdevelopments in this world shows that, while global living standards have risen significantly, this has been at the expense of natural resources: we already need 1.6 Earths to maintain the current standard. In the *Anthropocene*, the age in which humanity shapes everything, we seem to have forgotten that our lives and actions are embedded in the natural world that surrounds us, not existing outside of it. Biodiversity is disappearing faster than ever in human history—it is essential to achieve fundamental changes quickly. The Dasgupta study calls for a series of structural transformations, inter alia in industry and especially in food production, which must not lead to further destruction of the rainforest. Economic success must be measured not just in terms of growth and money, but also by indicators that evaluate natural capital and take a long-term perspective. Moreover, guidelines, protection measures and user fees must be introduced for ecosystems like the oceans, which are universal, public goods beyond the sphere of sovereign rights. For the preservation of ecosystems within national borders, the community of states should provide financial compensation. The transformation of the global economy and our use of natural resources, only hinted at here, will only succeed if the international community joins together to initiate profound systemic change without delay. Otherwise, the scale of environmental degradation and climate change will reach tipping points beyond which even more catastrophic consequences can be expected.[1]

The *Dasguta Review* refers to a looming scenario long since announced by the Intergovernmental Panel on Climate Change (IPCC): If we cannot change course immediately and drastically reduce greenhouse gas emissions, global warming will increase even more rapidly. The goal formulated at the 2015 Paris Climate Conference—to limit global temperature increase to 1.5° C—will probably not be achievable, and is likely to be accompanied by extreme weather events, heatwaves and floods, significantly affecting living conditions worldwide.[2]

Tourism in Climate Change

Tourism is affected by this development; and, like every other economic sector, it contributes to climate change and produces greenhouse gases. Tourism accounts for approximately five percent of all globally produced CO_2 emissions—and, unless massive countermeasures are taken, it will be ten percent *more* by 2030, the time-horizon for achieving the SDGs. Leisure travel accounts for about a quarter of this, with international tourism having a higher share due to air travel than domestic or interregional tourism. It is essential to reduce the emission of millions of tons of greenhouse gases and to decarbonize transport and tourism.[3] A step in this direction was taken by the Pan-European Programme for Transport, Health and Environment (PEP) with the Vienna Declaration. This was agreed in 2021, backed by the World Health Organization Europe (WHO) and the United Nations Economic Commission for Europe (UNECE) and includes measures for environmentally friendly and health-promoting mobility. It contains the first pan-European master plan for promoting cycling.[4]

Tourism, as a major economic sector with diverse impacts on social life and nature, is called upon to pursue and implement the Sustainable Development Goals propagated by the UN. The World Tourism Organization, established by the United Nations in 2003, attempts to network the objectives of the SDGs with tourism using a coordinated concept. This includes the Global Code of Ethics, the promotion of eco-tourism, and sustainable tourism for small island states. The Rio+20 process, the Istanbul Programme of Action on the use of marine and coastal resources, the Vienna Programme of Action for landlocked countries, and integration into the UN Habitat II Programme are further building blocks of this integrative strategy.

According to the UN-WTO, *sustainable tourism development* occurs when

- environmental resources are optimally utilized by maintaining essential ecological processes, contributing to the preservation of natural heritage and biodiversity:
- the socio-cultural authenticity of host communities is respected, their built and living cultural heritage and their traditional values are preserved and contribute to intercultural understanding and tolerance;
- viable, long-term economic operation is guaranteed: one that provides all participants with fairly distributed socio-economic benefits, including stable employment and income opportunities as well as social services for the host communities, and contributes to the fight against poverty.[5]

Among the action programmes for the protection of biodiversity, adaptation to climate change, containment of plastic waste and the propagation of clean energy forms for hotels, all based on examples of good practice, the

One Planet programme stands out in particular. It offers a road map for the tourism sector and its contribution to the SDGs, by building knowledge through successful examples, encouraging stakeholders to undertake their own projects and inspiring the implementation of new ideas and initiatives. Further, it enables the community of states to connect joint projects at private and public levels nationally and internationally. From these connections, learning experiences can emerge, which in turn find their way into national development plans and whose realization is documented in the voluntary national sustainability reports of the member states. This creates an international incentive system that promotes the achievement of the development goals.[6]

Turning Point in Tourism

The One Planet Sustainable Tourism Programme is implemented by the One Planet Network in cooperation with the governments of France and Spain and the United Nations Environment Programme (UNEP). The aim is to make the transition to a Green Travel and Tourism Economy: an ecologically oriented tourism economy. This vision, presented on *World Tourism Day* in June 2020, is also intended to aid the recovery of the tourism economy from the Covid-19 crisis and initiate a transformation of the entire industry — for better tourism, prosperity, and the planet.

The Covid crisis has made it clear that ecosystems can recover when economic activities and traffic partially or completely come to a standstill. The pandemic resulted in a seven percent reduction in greenhouse gases worldwide in 2020. Short-term emission pauses have no lasting effect, but such insights can raise awareness of the interrelationships in the population and strategically find their way into tourism projects — as in the form of proposals or rules for environmentally conscious travel. *Visit Valencia*, for example, has begun implementing an ambitious decarbonization program that is expected to lead to carbon neutrality of the destination by 2025. Another example is the *Visit Scotland* project, in which 5000 tourism companies/businesses focus on energy efficiency, avoiding food waste, and low CO_2 transport. *Intrepid-Travel*, a leading provider of sophisticated adventure travel, supports an innovative maritime permaculture project. This Australian company, with offices in Austria and Germany, is involved in the regeneration of seaweed forests in Tasmania, as algae remove CO_2 from the ocean, allowing the seawater to absorb more harmful CO_2 from the atmosphere. This is also an important step for complying with the recommendations of the IPCC and for achieving net-zero emissions by 2050.

The Covid-19 crisis has also sharpened awareness of regional production and supply chains, which play an important role in a green tourism economy. Circular value chains are a step towards sustainable economic growth because resources are used efficiently. Numerous pilot projects are underway to avoid food waste. The recycling of plastic waste is of massive importance because 13 million tons of plastic end up in the oceanseas every year, to the detriment of the fishing and tourism industry. The French government supports a large project, backed by over 60 tourism organizations, that recycles plastic waste for circular reuse. Argentina is leading a multilateral initiative aimed at combating climate-damaging forms of food production and also trying to steer the production of plastic into sustainable channels. Also on board are the European Commission and waste-separating countries such as

Germany, the Netherlands, Sweden, and Switzerland, as well as Bhutan, Costa Rica, and Mauritius.

Communication and monitoring of such examples are prerequisites if tourism with ecological and social insight is to become a viable model for the entire industry and for greater public awareness of the complex interrelationships.

This is also a question about green jobs in tourism, social inclusion, and fair wages for employees, as well as supporting micro-enterprises, which can guarantee high diversity for the industry. Canada, for example, supported indigenous tourism micro-businesses during the pandemic with a stimulus programme. In this way, the local labor market could be relieved and the businesses could use the time to adapt to the latest market developments. New connections with the creative industry were also established, and there were closer contacts with local communities. Strengthening local value chains ultimately makes them less dependent on external suppliers—and the local tourism economy benefits from domestic tourism.[7] At a meeting of the G20 tourism ministers in May 2021, they committed to the digital and green transformation of the tourism economy.

Global tourism was increasingly resource- and emissions-intensive until the outbreak of the covid pandemic. Indeed, it was heavily criticized in a joint statement by TourCert, TourismWatch, Bread for the World, and Studienkreis Tourismus & Entwicklung. The tourism-critical NGOs demanded swift implementation of the Agenda 2030 in tourism, to combat the negative ecological and social impacts of tourism. As sustainability is linked to economic development, ecological and social justice, the goal must be fair and responsible tourism that conserves natural resources, inspires travellers and employees for sustainable action, and conveys joy, nature experiences, a sense of community, and the desire for a sustainable lifestyle. This must form the basis of any tourist development. Such a tourism turnaround must also entail questioning existing models and halting climate-damaging practices such as the promotion of high-emission forms of mobility. We must legally anchor, implement, and create incentives for consistent protection measures for people and the environment, so that all actors, including tourism companies and travellers, contribute to sustainable development.[8]

Searching for Ways to Sustainability

Starting from the World Summit on Environment and Development in Rio 1992, a phase of reflection on the permanent protection of resources was initiated. The UN-WTO sees Sustainable Tourism as that form of economic development that leads to an increase in the quality of life in the host countries, offers a high quality of experience for the visitor, and at the same time contributes to the long-term preservation of the environment.

For years now, the UN-WTO and numerous civil society institutions have been presenting programmes, recommendations, declarations, resolutions, and guidelines for sustainable tourism development—but these have been largely ignored by tourism providers, or implemented very hesitantly. In the UN General Assembly in December 2020, the significance of a global sustainable tourism development was once again communicated as a resolution to the world public.[9]

Referring to the SDGs such as poverty reduction or improving living conditions for millions, the UN-WTO has been advocating goals and measures, in the face of impending climate change, to align present and future tourism with sustainability. As many of the following suggestions or indicators as possible should be pursued to achieve the sustainability goals:

- Focus on long-term economic profitability (competence, attractive destinations promote customer loyalty)
- Maximization of regional prosperity (minimal outflow of revenue as possible, creation of value chains)
- Quality of employment (year-round jobs, appropriate pay and working conditions, good training, employees recognized as the capital of tourism businesses)
- Fair distribution of benefits and social balance (fair distribution of income from tourism)
- Visitor satisfaction (quality of service, price/performance ratio)
- Participation in future development (control at local level, involvement of the next generation)
- Improvement of local quality of life (tourism intensity, not overburdening capacity, information work)
- Respect and strengthening of cultural heritage (preservation of built heritage, sensitive handling of intangible heritage)
- Protection of the environment, ensuring the quality of living space (minimal destruction of the landscape; appropriate infrastructure for tourists and locals)
- Protection of biodiversity (cooperation with national parks etc., promotion of nature-based tourism)
- Efficient use of natural resources (minimal consumption of water and non-renewable energy, careful tourism planning)
- Minimization of emissions and waste production (gentle mobility, adapted technologies).

The goal of *sustainable tourism* is achieved when the following criteria are met equally:

- Tourism is possible in the long term, because the development or use of all resources is conducted gently
- is culturally compatible, with for local conventions and rites, renunciation of exploitative commercialization, and adaptation to local standards
- is socially balanced: the benefits and disadvantages are equally distributed, regional disparities are avoided, and local actors are significantly involved in the decisions
- is ecologically sustainable, because there is minimal pressure on the environment, with avoidance of damage to biodiversity and promotion of environmental awareness
- is economically sensible and productive, because it is a profitable business for the local or national economy and contributes significantly to the creation of income for the local population. I have called this the *Pentagon of sustainable tourism*.[10]

The reports of the Intergovernmental Panel on Climate Change (IPCC), which provide evidence on the basis of thousands of studies that climate change is due largely to human intervention in ecological systems, have at least created uncertainty throughout the tourism industry—especially since the changes in the natural structure also indicate significant problems for the tourism economy.

The industry's reactions have varied greatly, ranging from very serious efforts to green-washing. The whole contradiction in the actions of tourism providers is seen in the following example: The TUI Group, which, in the 1980s, was the first travel company to appoint an environmental officer, now operates a fleet of 16 cruise ships. Only the latest generation is equipped with liquid natural gas propulsion and does not run on heavy oil. Marine residual oil is the trade name for the pollutant-rich residues from oil processing, which mixed with diesel oil as ship fuel pollute the oceans. Despite constant improvement of energy efficiency, the TUI fleet "Mein Schiff" consumed 141,000 tons of heavy oil in 2019 and burdened the environment with 475,000 tons of CO_2 emissions.[11]

Compared to one billion tons of carbon dioxide from global shipping, about three percent of all human-caused emissions (plus 15% of global nitrogen oxide emissions and 13% of sulfur dioxide emissions, still rising)[12], this consumption for the sake of pleasure may not seem excessive. However, TUI had a share of some three percent of the market of 32 million cruise passengers in 2018.[13]

The fact that cruise ships produce high amounts of greenhouse gases is not new. Even though shipping companies may be interested in increasing energy efficiency for economic reasons, many tourism providers still do not feel responsible for the decarbonization of travel.

Air travel, whose fossil fuel is even state-subsidized in many countries, also generates a massive ecological footprint. In 2019, there were approximately 47 million flights worldwide, and total air travel is responsible for 3% of global emissions from the combustion of fossil fuels. The airplane is the mode of transport that puts the most strain on the environment per passenger kilometer. Environmentalists in Germany and Austria have therefore been calling for the introduction of a aviation fuel tax and a ban on domestic flights. In Germany, the federal government increased the air travel tax at the end of 2019 to make flight tickets more expensive, citing reasons of climate protection.[14]

Voluntary compensation, where passengers accept a surcharge on the ticket depending on the distance and CO_2 emissions involved, has not yet been widely adopted. With such compensation payments to the German NGO *atmosfair* (in 2019, ca. 20 million euros), climate protection projects of this organization concern countries of the global South.[15]

Climate change has now become one of the most urgent concerns of Europeans. Many are therefore in favor of climate-friendly travel, but attitudes are not always reflected in actual behaviour. After all, who wouldn't like to fly cheaply to Mallorca, or enjoy a full English breakfast in London? 'Flight shame' has become a public issue in the context of the *Friday for Future* movement, on the other hand the Swiss voters rejected an increase in car fuel prices and an increase in air fares in a referendum. As long as the aviation industry cannot rely on *Sustainable Aviation Fuel*, flying will not become more climate-friendly. Moreover, only one percent of the world's population is responsible for 50% of emissions in air travel, but leaves an enormously high footprint.[16]

With the Framework Convention on Climate Change signed in Rio in 1992, the signatory states committed themselves to reducing the amount of greenhouse gases emitted worldwide. The Kyoto Protocol requested states to reduce their emissions by an average of five percent annually, but they have been lagging behind these goals. State or supranational steering measures are particularly important because they accelerate the process of introducing climate protection measures through regulations and funding policy. But every tourism region, every tourism business and every tourist can also make a contribution.

Dimensions of Responsibility

The severity of the storms in the summer of 2021 has also made the urgency of rethinking clear to the tourism industry. Flood disasters in Germany and Austria, hailstones the size of golf balls, a tornado over the Czech Republic, heatwaves and forest fires in the Mediterranean countries; the threats have arrived at our own doorsteps. Climate change and the sustainability dimension have been making the headlines and talk-rounds on television—a good 40 years after the first studies on the proven human-induced climate change and the call to treat the planet more gently and carefully.

For tourism, this re-orientation means recognizing the complexity of sustainability goals and implementing them at the national, regional, and local levels in the form of measures and projects. Christian Baumgartner[17] divides this task into *dimensions of responsibility*. Tourism is to be understood as an integral part of a sustainable, region-specifically networked economy *(economic dimension)*. An intact natural or living environment and operational environmental protection are necessary prerequisites for the tourism of the future *(ecological dimension)*. Holiday regions have a right to self-determined cultural dynamics, the population and those employed in tourism have a claim to social satisfaction *(sociocultural dimension)*. Intensively used tourist destinations must develop and apply operational and municipal environmental management systems as well as local sustainability strategies. The population must be an equal partner in shaping tourism policy, involved in decision-making processes. The tourism source-areas of the conurbations and higher-level political systems share responsibility for the effects of tourism in the holiday regions *(institutional dimension)*. People seek a change of location to compensate for local shortcomings, seeking places with subjectively higher quality of life and experience *(individual dimension)*. Finally, the type of leisure mobility largely depends on the service offers or alternatives (arrival packages, door-to-door service, etc.) and requires regions to control mobility through appropriate offers *(service dimension)*.

Politics, tourism and production companies, civil society institutions, and the science are the main actors in this transformation process — they must create the basis for travellers to behave ecologically. Through their interaction and the holistic consideration of local conditions, we must find lasting solutions for regional development in balance—because sustainability not only aims at preserving natural resources, but also strives for lasting stability of ecological, economic, and sociocultural conditions.

A good example is provided by the *Cheese Road Bregenzerwald* in Vorarlberg, the westernmost federal state of Austria. From the merger of farmers, alpine dairymen, and cheesemakers, cheese innkeepers and guesthouses, museums and local railways, tourism businesses, and partners from crafts and trade, a 'brand' was created for an entire region.[18] Another example of good practice of an integrative strategy is the *World Heritage Cinque Terre* in Italy. This terraced cultural landscape, a World Heritage site since 1997 and a national park before that, offers a unique *terroir* for grapes, lemons, olives, and herbs. The preservation of traditional cultivation methods and the entire agricultural system was achieved not least through the protective measures and the management system implemented with the award of the World Heritage title. Investments were linked to a focus on high-quality tourism and the creation of a protected origin brand that corresponds with high quality and appropriate prices — a central element in this highly competitive market.[19]

Cinque Terre now has a prominent place in the initiative of the European World Heritage Vineyards initiative, which also cooperates with the World Heritage Wachau in Lower Austria. This region too is distinguished by UNESCO, because the tangible material testimonies of the cultural landscape in this stretch of land along the Danube river, not far from Vienna, have been preserved to a remarkable extent, in a historical development lasting more than two thousand years. Here too, the protection of cultural assets in connection with the production of outstanding products, their geographical indication, and high tourist attractiveness, guarantees a considerable economic performance and thus the valorization of the cultural landscape.[20]

Nico Stehr—Founding Director of the European Centre for Sustainability Research—speaks in the context of sustainable development of the demand for a "responsible growth of the economy" and observes a "moralization of the markets". This manifests itself not only in changes in the behaviour and orientations of market participants, but also in the products and services traded on markets, thereby obliging producers, supply chain, and consumers to act responsibly.[21]

Mobility to and at the Holiday Destination

To make tourism more climate-friendly, we must start with transport, with the traffic to and from holiday destinations.[22] According to studies by the Austrian Panel on Climate Change, car travel within Austria or from neighbouring countries to the mountain and lake regions of the Alpine Republic, is up to 80%.[23] This also applies to many holiday trips to the Mediterranean coasts, and leads to endless traffic jams during the holiday season. Individual car traffic is also a trigger of *overtourism*.

The Alpine region, a high quality natural living and tourism area, is characterised by high car mobility due to its peripheral location. The demand-oriented design of sustainable mobility offers and the provision of environmentally-friendly means of transport for tourists and locals are of central importance for a climate-friendly regional policy. Some successful pilot attempts in the Alpine regions exemplify how smart mobility, continuous CO_2-low travel chains for arrival and departure, and multimodality at the holiday destination can be practised.

In the Tyrolean Ötztal Alps, public transport and travel by train are promoted. With the initiative *Tyrol on Rail*, Tyrol Advertising office has been pushing for a coordinated mobility chain. A pilot project *Easy Travel* by the Ötztal Transport Company with the University of Innsbruck has now become a permanent offer. Guests arrive by train, a pick-up service takes them and their luggage to their hotels in the five municipalities. Around one million guests, who generated over four million overnight stays in the calendar year 2019, came to the 65 km long valley — but 85% of them by their own car, as travel habits do not change that quickly. There is an enormous potential for emission savings when switching to public transport.[24]

Customer surveys have shown that problems of luggage transport, the need to change trains, and longer travel times often speak against travelling by train. A better luggage service, shuttle services, and the free use of public mobility offers at the holiday destination are desired. That mobility does not automatically mean *private car* mobility can be demonstrated by the Vorarlberg project *Mission Zero Silbertal*, where multimodal offers were created for winter and summer tourists, including ski and hiking buses, a guest shuttle, car-sharing offers, and a call-up collective taxi.

Similarly, the Austrian Verkehrsclub (Traffic Club VCÖ) a few years ago: as part of a regional mobility offer to make private cars superfluous the digital travel platform of the tourism region Villach-Faker See-Ossiacher See arranged for arrival by train, pick-up, luggage service, and the provision of further means of transport. This was the second major attempt to promote soft and multimodal mobility, following the model of the municipality of Werfenweng in Salzburg's Pongau region.[25]

Also tourists who arrive by their own cars can manage without, if appropriate mobility offers on site exist. This can easily be done via regional guest cards with included use of public transport or mobility hubs. If mobility offers are simple, reliable, comfortable, and clear, then they will be accepted and long-established travel habits will gradually change. In general, central platforms for information/booking for arrival and departure should include as many multimodal offers as possible to promote continuous travel chains and information regarding the best offers and additional services (e.g., luggage service, hotel shuttles). Changes should be kept as minimal and customer-friendly as possible. It is also important to make public transport bicycle-friendly at all times, to provide transparent information, and to introduce or optimize combination ticketing.

The example of the *Vinschger Bahn* in South Tyrol/Alto Adige shows how public transport offers for locals and tourists can contribute to improving the quality of life and experience.[26] This railway connection between Mals and Meran had to be constantly expanded due to high demand from tourists and locals. New offers were created for tourists throughout South Tyrol. These include guest cards, apps for online booking, the modernization and electrification of the Meran–Bozen route, and connecting the ski areas in the Puster Valley. The train is used in roughly equal parts by seniors, students or commuters, and tourists, with a huge increase in cycle tourism. Similar plans exist in the Salzburg region with the Pinzgau Bahn and are partly already being implemented.[27]

The project *Stop 4.0* by the Research Studio iSPACE takes into account the increasing importance of cycling, not only in the tourist context. Under the slogan *region-specific personal mobility*, the pilot project cooperates with the Salzburg State Transport Organisation and is looking for solutions for commuters and tourists. A data laboratory is being created at Neumarkt/Wallersee station and a concept for using existing infrastructure for cycle tourism in the Salzburg Lake District is being developed.[28]

Until the outbreak of the pandemic, World Heritage Sites and historic old towns across Europe were overrun by cars and hit-and-run tourism. However, through social media applications — available via digital information platforms — visitor flows can be traced and thus also spread out, preferably even controlled.[29] This benefits the quality of the tourist experience as well as the quality of life of the locals. Unfortunately, many destinations, such as the Pragser Wildsee/Lago di Braia in the Dolomites, which has been overrun by Instagram-driven tourists for years, are still waiting for a truly problem-solving visitor management system. However, the digital management of coach traffic for the world heritage sites of Salzburg or Hallstatt has made things easier.

New information technologies can be used more extensively in the future to optimize mobility and provide policymakers with decision-making aids. They can reveal the interrelationships between traffic, settlement, and energy, as offer-oriented traffic planning is necessary to reduce greenhouse gases and CO_2-emissions. This is in line with the demands of the EU catalogue of measures and the Paris Climate Agreement. They must guide actions if the intended goals are to be achieved.

The European Union's *Green Deal*, a law passed by the EU heads of state and government, stipulates that Europe must be de facto emissions-free by 2050. By 2030, as an interim goal, 55% of all greenhouse gases must be saved (*Fit for 55 package*). This is to be achieved through a mix of market-based instruments like emissions trading and legislative requirements such as a CO_2 import tax and emission limits for motor vehicles, which will be almost exclusively electrically powered from 2035 onwards. To achieve such carbon neutrality, the entire economy and all transport technologies will have to be transformed. Hundreds of billions of euros are earmarked through funds, subsidies, and innovation aids to kick-start and cushion this transformation. The tourism industry, like all other economic sectors, is called upon to follow this transformation process — to save energy, convert heating systems to renewable energy, adapt businesses to the challenges of climate neutrality through thermal insulation, and much more. Important here is more environmentally friendly transport, where emissions must be reduced by 90% by 2050.[30]

Visions and Measures

In Austria, the general demands of the sustainability discourse have been taken up in documents such as the National Environmental Plan and formulated as objectives and framework conditions at regional and local level. In the field of tourism, the 5th Action Programme of the EU and the protocols of the Alpine Convention include regulations for a sustainable environmentally friendly development. In 1995, the Austrian government adopted principles and goals for a sustainable tourism and leisure industry. However, day-to-day practice and the implementation of measures have lagged far behind these goals—and not only in Austria. Even though the principles of a sustainable tourism and leisure economy have been operationalized in some regions through eco-audits, they are still little used outside of protected areas. Especially in the Alps, such monitoring should be part of leisure economic activities.

Like Germany and Switzerland, Austria also had to present a climate protection strategy to compensate for the implementation failures of previous governments. Whereas most EU countries managed to reduce their CO_2 emissions from 1990 to 2017, they increased in Austria by five percent, due mainly to traffic.[31] Milestones of a climate-neutral policy are the recently passed Renewable Energy Sources Act (EAG), which aims to ensure that only 'green' electricity will be coming out of sockets from 2030, and a comprehensive expansion package for the Austrian Federal Railways. The 1-2-3 climate ticket, a low-cost annual pass for rail passengers, was recently introduced, as was the eco-social tax reform announced in the government programme, which includes CO_2 pricing.[32]

In this context, *spatial planning decisions* are important, as they involve long-term responsibility for climate change and cultural and living space in connection with transport policy. Road expansion, the construction of new tourist infrastructure and the construction of further chalet villages or the re-zoning of green land into building land for tourism businesses or shopping centres have received considerable public attention, heavily criticized with reference to climate change.[33]

In Austria, regional tourism associations play a key role as shapers of domestic tourism. They are entrusted not only with the creation of marketing concepts, guest care, or tourism facilities, but also explicitly with the promotion and preservation of culture and landscape. This can prove problematic because economic interests usually dominate the decisions made in tourism associations.

The *Tourism Protocol of the Alpine Convention*, on the other hand, is dedicated to holistic development and the protection of the Alpine region. It states that tourism is in the public interest and that this sector offers a chance of survival for many regions. The signatory states commit themselves to contribute to environmentally friendly tourism and sustainable development of the Alpine region, in the areas of spatial planning, transport, agriculture and forestry, environmental and nature conservation, and water and energy supply.[34]

Plan T—The Plan of Strong Headlines as a Case Study

In 2019, the Austrian government presented its Plan T as a guideline for Austrian tourism in the future. Although it is little more than an instrument for economic promotion, the master plan is entitled "On the way to the most sustainable tourism destination"—a gross overestimation.

The plan was developed through surveys of 600 companies, 30 stakeholder interviews and nine future workshops, involving a total of 500 people—a huge effort and certainly good for the internal climate of the industry. But what is not listed here is anything really new or surprising. Most of this should have happened a long time ago. At least climate change is now officially recognised as a challenge.

As to *sustainability*, however, we find only vague declarations of intent, with some points that have been known for years—such as arrivals and departures 80% with private cars, and that CO_2 emissions must be reduced by

36%. Tourism must recognise its global responsibility and contribute to the implementation of the UN Sustainability Goals. But how? The more concretely the problem is presented, the more tight-lipped do the answers become. Apparently, it is not popular to impose too much actual change on the industry.

There is a lot of talk about 'intact nature', but the landscape is already scattered and concreted, perhaps more so in Austria than elsewhere in Europe: spatial planning laws are toothless, apartment blocks and chalet villages are growing rapidly, to the detriment of the local population around the settlement centres, making living space scarce and expensive. This misdevelopment also damages tourism. Importantly, there is no mention of the intention to contribute to the preservation of the remaining cultural landscape: this would mean defining the regulatory and spatial policy framework to protect overburdened terrain from further human intervention. The basis of life must be secured sustainably. Tourism should take place in harmony with people and nature. We must reduce greenhouse gas emissions, rely on renewable energy, increase resource efficiency—and maintain the competitiveness of the tourism location. These are demands on domestic tourism that have existed for several decades now. Yes, it is laudable to pick up on this now, encouraging the industry to increase its commitment to the '100,000 roofs photovoltaic and small storage programme' and other funding measures. But this will not be nearly enough.

In the Austrian Alps and elsewhere, the expansion of sustainable mobility and transport solutions is being pursued, priorities are being set in urban as well as rural regions—but why has it taken so long to realise that things cannot continue as they have been? Arrival and departure as well as local mobility should be designed to be climate-friendly; likewise, better networking of the tourism and transport industry, and, integrative transport systems are desired — but what is actually planned? There has been no mention of expanding public transport systems or of innovative solutions to modal split, nor of slowing down the development of alpine regions and glacier areas through new cable car solutions.

Sustainability must be established as a unique selling point for Austrian tourism. This might include making the tourism industry itself a power plant, with businesses generating and supplying their own renewable electricity. There is enormous potential here, and also the possibility that hotels can become energy self-suppliers.

Much of the Austrian government's Plan T resembles a declaration of intent to boost the economic support of the industry. This is also stated in a subheading: The focus must be on *value creation*. The reduction of sales tax as a partial success is mentioned, further measures are desired, administrative simplifications are demanded—especially help and advice on the transfer of some 2000 hotels to the next generation. Why this is such a problem in this industry? Does it have to do with the commitment, the working model, the profile of requirements for entrepreneurs and employees, the hardships of the service industry? It is demanded from AirBnB accomodation providers of the sharing economy that they should pay taxes like any private person who rents out rooms. This reduces competitive distortions and has already been implemented by the responsible state governments. New financing instruments are desired from the Austrian hotel bank, and will also be provided. This shows how much the government cares about tourism and gives importance to the sector. During the Covid pandemic, the Austrian government compensated the loss of turnover of the hotel and restaurant businesses by 80%, including every luxury hotel; it granted tax deferrals and a short-time work programme for the employees.

A hot issue in this Plan T is the staffing problem. The search for personnel is becoming increasingly difficult, also due to the success of the industry. *Alpine Culinary* is supposed to become a focus in the coming years, but how can that be, given the shortage of hundreds of chefs even before the pandemic? Tourism offers attractive opportunities for young people — but the young people apparently do not see it quite like that: unfilled apprenticeships are still to be found in tourism. The goal is to increase the attractiveness of working in tourism—does this also mean higher wages, regulated working hours, adequate accommodation and better personnel management? Like many other industries, tourism needs more immigration, but simply pointing to the control of the official immigration card, the *red-white-red card*, will not solve this problem, as more skilled workers are needed from abroad. In this respect, the Austrian government's restrictive immigration policy—by sending successful apprentices back to Afghanistan, Syria (etc.) immediately after their training—is a self-defeating policy.

Farm holidays have been a cash cow of Austrian tourism for years, but to call this an example of success for the balance between tourist and agricultural and forestry use seems exaggerated. Farmers in the alpine regions have always needed additional income to make ends meet. Tourism undoubtedly plays an important role, as it provides decentralized jobs or additional income. Closing value chains through regional products is also important and has been pursued for years. However, innovative ideas and visions for the future are missing: it is important to go beyond what already exists. Expanding this cooperation is certainly advantageous for all parties involved.

From destination marketing, the aim is to move towards destination management — the management of living spaces for locals and guests. This will challenge regional or local politics, as not everywhere do the local populations see themselves as service providers; many have a quite critical view of tourism. But the markets are to be conquered, together — especially the growing ones in Asia (again, air travellers); and in Central and Eastern European countries, the Germans are the most important market and must be kept on board, as well as the Austrians. This takes us back to marketing. Growth still remains the measure of all things, and in the action plan for 2019/20, special budgets are planned for the marketing campaigns carried out by the Austrian National Tourist Office.

Accentuated by the pandemic and the disasters of the summer of 2021, the mild winds of Plan T are likely to bring substantial changes. Willingness to engage seriously with the issue of sustainability has become more evident in some tourism associations, hotel management, in the transport sector and in the trade with holiday, leisure and sports articles. Of course, the aim is to make tourism a profitable growth industry again, to pick up as quickly as possible where it left off in Spring 2020. There are indeed some significant changes underway in Austrian tourism. This includes a gradual re-orientation towards local markets as well as a focus on higher quality standards, the merging of destinations into experience regions, in communication and marketing the increased digital approach to guests—as well as new offers such as 'workation' stays, combining work and vacation. With the emphasis on the regional and its underlying value basis, guests are invited to share these with the locals and to enjoy the high quality of life. There are also new climate-friendly mobility offers, to increase the quality of visitor stays and transform town centres into low-traffic zones. Cable car companies have started sustainability processes, and a growing number of mission statements and tourism strategy papers express the desire for harmonization of tourist economic interests and the needs of the local population.³⁵

Could You Hear Me?

When, in 2006, I gave the first lectures aimed at drawing attention to the massive problems of future winter tourism due to climate change, backed by solid data from the World Climate Council and the research group which the Swiss government had set up for advice, I received little positive feedback from the industry. In that year, the first extensive Alpine snowfall came only in the middle of the winter season, and the nerves of all those who live from winter tourism were already on edge when I appeared at the invitation of the Green Party's parliamentary group in Zell am See. Hoteliers, tourism directors and managing directors of cable car companies already considered the title "Sweating Planet and Cold Beds" a malicious provocation, and anyone who openly criticizes tourism in Austria was immediately stamped as a business-damaging subject. And indeed, my warnings simply bounced off—the industry didn't want to hear anything about global warming; they saw me and all climate researchers as charlatans, fearmongers or eco-fundamentalists. The powerful president of the Austrian Ski Association, himself a major cable car and tourism entrepreneur, thought it was all a scam. In an interview, he told the media that he had read a book that questioned all the data from the Intergovernmental Panel on Climate Change and therefore climate change itself.

Today, we are *almost* literally up to our necks in water, and tourism is fighting for its existence. Like many politicians worldwide, decision-makers in tourism have done surprisingly little to address the predicted effects of climate change and make the tourism system more resilient to comprehensive changes. I personally believe that there is no alternative but to try to implement the *Pentagon of sustainable tourism* as explained earlier in this text. The chances are good, because in the face of the harbingers of the apocalypse, global politics and the European Union in particular seem ready to act quickly with the available means and with the help of nascent technologies to achieve long-term objectives.

The warning against short-term thinking fits perfectly with trees and their slow but steady growth; indeed the idea of sustainability can be traced back to forestry: the careful use of resources so that benefits are continuously created, there is always enough wood available, with adequate re-growth ensured.[36] Now the forests are also in danger: due to climate change, some types of trees can no longer tolerate our latitudes, because we humans have interfered too much with the system's self-regulation. The fight for every tenth of a degree less global warming must begin immediately.

Endnotes

1. https://www.gov.uk/government/publications/final-report-the-economics-of-biodiversity-the-dasgupta-review; accessed 9.8.2021.
2. https://www.de-ipcc.de/media/content/Hauptaussagen_AR6-WGI.pdf, accessed 10.8.2021.
3. UNWTO: Transport-related CO_2 emissions of the Tourism Sector. https://www.unwto.org/sustainable-development/tourism-emissions-climate-change, accessed 7.7.2021.
4. https://thepep.unece.org/sites/default/files/2021-03/2103410E.pdf

5. https://www.unwto.org/unwto-un-system; https://www.unwto.org/sustainable-development
6. See Tourism and the Sustainable Development Goals—Journey to 2030, https://www.e-unwto.org/doi/epdf/https://doi.org/10.18111/9789284419401.
7. https://webunwto.s3.eu-west-1.amazonaws.com/s3fs-public/2021-05/210504-Recommendations-for-the-Transition-to-a-Green-Travel-and-Tourism-Economy.pdf?wiwmhlGgXT4zwXles_Q8ycdITGIQfaMt
8. Tourism turnaround—Agenda 2030 for sustainable development: Shaping the transformation in tourism (2016); https://www.tourism-watch.de/system/files/document/Profil20-de-v07-Web.pdf
9. https://digitallibrary.un.org/record/3879234; https://undocs.org/en/A/RES/75/229.
10. Kurt Luger, Schwitzender Planet—kalte Betten (Sweating Planet—Cold Beds. Tourism in the age of moralizing markets and climate change), in: Roman Egger & Thomas Herdin (eds.), *Tourismus Herausforderung Zukunft* (Tourism Challenge Future), LIT: Vienna 2007, 127–142.
11. https://www.tuigroup.com/de-de/verantwortung/engagement/kreuzfahrten, accessed 10.8.2021.
12. The world fleet of around 90,000 ships burns some 370 million tons of fuel per year and also produces 20 million tons of sulfur oxide. https://www.srf.ch/kultur/wissen/schifffahrt-das-schmutzigste-gewerbe-der-welt, accessed 10.8.2021.
13. TUI also relies on ships from other shipping companies, such as Costa Cruises. https://de.statista.com/statistik/daten/studie/285194/umfrage/passagiere-auf-dem-weltweiten-kreuzfahrtmarkt/;Study_id55329_weltweiter-kreuzfahrtmarkt.pdf, accessed 10.8.2021.
14. https://de.statista.com/statistik/studie/id/70799/dokument/flugpassagierverkehr-und-klimaschutz/, accessed 10.8.2021.
15. https://www.atmosfair.de/wp-content/uploads/atmosfair_2019_jahresbericht_deutsch.pdf, accessed

10.8.2021; for measuring your own footprint see www.fussabrucksrechner.at of TU Graz.
16. European Commission, Standard Eurobarometer 92, Autumn 2019. Public opinion in the European Union. Survey November 2019. https://ec.europa.eu/commfrontoffice/publicopinion/ accessed 30.10.2020; *Reisewarnungen* (Travel warnings. Portfolio Freizeit und Leben), in: Profil, August 8, 2021.
17. Christian Baumgartner, *Nachhaltigkeit im Tourismus* (Sustainability in Tourism). Innsbruck: StudienVerlag 2008.
18. https://www.kaesestrasse.at/
19. CHCfE Consortium, *Cultural Heritage Counts for Europe*. Krakow 2015. https://www.digitalmeetsculture.net/article/cultural-heritage-counts-for-europe-final-report/ accessed 02.04.2019; Sarah May, *Ausgezeichnet!* (Excellent! On the constitution of cultural property through geographical indications.) Göttingen Studies on Cultural Property, Göttingen: Universitätsverlag 2016.
20. Verein Welterbegemeinden Wachau (Association of World Heritage Communities Wachau), *Wachau World Heritage Management Plan*, Vienna 2017.
21. Nico Stehr, *Moralisierung der Märkte* (The Moralization of the Markets. A Social Theory). Frankfurt: Suhrkamp 2007.
22. See also Hartmut Rein & Wolfgang Strasdas (eds), *Nachhaltiger Tourismus* (Sustainable Tourism), Konstanz: UTB 2015.
23. https://ccca.ac.at/wissenstransfer/apcc/broschuere-der-oesterreichische-tourismus-im-klimawandel, accessed 10.8.2021.
24. https://www.unibk.ac.at/newsroom/easytravel-ohne-auto-in-den-urlaub.html.de, accessed 10.8.2021.
25. https://connect.visitvillach.at/de/das-mobilitaetsangebot-der-region.html
26. https://www.sta.bz.it/de/eisenbahnen-seilbahnen/vinschger-bahn/

27. Kurt Luger, Mobilität zum und im Urlaubsort (Mobility to and at the holiday destination), in: *Forum Mobil,* 4/2020, 14–16.
28. https://www.researchstudio.at/projekt/uml-haltestelle-4-0/, 30.8.2021.
29. Engelbert Ruoss & Angela Sormaz, Social Media and ICT Tools for Managing Tourism in Heritage Destinations, in: Kurt Luger & Matthias Ripp (eds), *World Heritage, Place Making and Sustainable Tourism,* Innsbruck: StudienVerlag 2021, 247–270.
30. https://ec.europa.eu/info/strategy/priorities-2019–2024/european-green-deal_de, accessed 12.8.2021.
31. https://www.bmlrt.gv.at/umwelt/klimaschutz/klimapolitik_national/anpassungsstrategie/strategie-kontext.html, accessed 10.8.2021.
32. Astrid Gühnemann, Agnes Kurzweil & Markus Mailer, Tourism mobility and climate change: A review of the situation in Austria, in: *Journal of Outdoor Recreation and Tourism* 2021. https://doi.org/https://doi.org/10.1016/j.jort.2021.100382.
33. The gaps in spatial planning laws, which make the sale of the homeland possible in the first place, are brilliantly analysed by Franz Dollinger, in: *Das Dilemma und die Paradoxien der Raumplanung* (The Dilemma and Paradoxes of Spatial Planning), Vienna: LIT 2021.
34. https://www.cipra.org/de/cipra/oesterreich/das-tourismusprotokoll-der-alpenkonvention-bedeutung-und-anwendung
35. Neuausrichtung (Reorientation), in: *Bulletin,* magazine published by the National Austrian Tourism Office, Issue 2/2021, 18–27.
36. Frank Uekötter, *Im Strudel, Eine Umweltgeschichte der modernen Welt* (In the Whirlpool, An Environmental History of the Modern World), Frankfurt: Campus 2020.

10

The Vision: Smart Tourists, Minimally Invasive

When you go to another country,
you should know what is forbidden there.
Confucius

Take nothing but pictures, leave nothing but footprints.
Sir Edmund Hillary

It is a late afternoon in the monsoon season, and the sun finding its way through the clouds. The fields around the village of Bulung, on Nepal's north-eastern border with Tibet, are steaming after the rain, the snow-capped peaks of the surrounding Himalayan giants glittering like holograms. With EcoHimal staff, I walk from village to village, checking on the progress of the Rolwaling Ecotourism project. Ahead of the group, I reach a small hamlet, which with its lush subtropical vegetation looks like a picture postcard: women in colourful dresses and decorated with brass jewellery are weaving cloth, men with broad knives at their belts are weaving baskets, infants are sleeping in rocking cradles, cattle and goats are grazing, water buffalo are wallowing in the pond, a hen with her chicks is out hunting for maize grains or juicy leeches, and a few dogs are dozing off. I would have loved to photograph this seemingly idyllic scene, but the dogs barked and suddenly all eyes—human and animal—turned in my direction. But this peaceful idyll still exists in front of my eyes and is repeated every day in the thousands of villages of Nepal: this is everyday life, nothing exotic or romantic. The subsistence farmers of Nepal have to work hard to wrest a livelihood from this challenging land to get their families through the year.

We destroy what we seek by finding it — in this case a picture-perfect situation of apparent harmony between man, animal, and nature but how far does the destruction or impairment caused by tourists go? What is actually affected, and to what extent?

L'idiot Du Voyage — The Naive Traveller

Too many tourists and travellers are unaware that they are part of a potential problem. On the contrary, they like to think that their presence and what they spend on accommodation and food, what they buy as souvenirs, deserve the gratitude of the local population: after all, don't they bring development and prosperity to the peripheral regions? They are generally unaware of the connections and effects of their presence.[1] Jean-Didier Urbain, in his ethnographic analysis, dealt with the figure of the tourist, who has to find his way in a foreign country with little skill and much ridicule. Not every traveller succeeds in his quest for a worldview, and without a real goal and adequate preparation for the foreign place, the venture must remain fruitless. Urbain's journey into the world of the tourist shows us that today's clichés have a long tradition and that the mythical innocence of the traveller may never have existed.

Poorly informed, insensitive tourists flock to the Alps. They try to walk across glaciers in flip-flops, drive on avalanche slopes, and meet death with the snow slabs they trigger. National park rangers have noted how tourists like to feed marmots with chocolate bars. Also tourists trekking in the shadow of the Himalayan eight-thousanders see no problem in handing out sweets to begging children—in an area without dental care system. Pencils would be more sensible.

Looking back in history, we can see how would-be mountaineers came from the cities to the countryside, but found that they could not get by without the skills and help of local alpine farmers. They had to learn the skills of the peasant chamois hunters. This required a certain cultural symbiosis. It took a century for the bourgeois alpinists to acquire the ability to climb without a guide.[2] The principle of self-responsibility inherent in this remains valid in alpinism today, where good preparation with precise route planning, appropriate equipment and good physical condition, as well as reliable weather forecasts, are prerequisites for reaching the goal and making a safe return.

This fundamental cooperation between travellers and those being visited is also referred to in the UNWTO *Code of Ethics*.[3] It speaks of mutual understanding and respect between peoples and societies, of ethical values common to humanity, tolerance and respect for the diversity of religious, philosophical, and moral beliefs—all of which may be promoted by tourism. Tourist activities should be conducted in harmony with the characteristics and traditions of the host regions and countries, and in compliance with their laws, practices, and customs. Indeed, tourists and visitors have a duty to familiarize themselves with the specific characteristics of the countries they wish to visit—before they set out. They need to be aware of the health and safety risks associated with any trip outside their usual environment, and should behave in such a way that these risks are kept as low as possible. It goes without saying that they should not engage in actions deemed criminal under the laws of the visited country, and refrain from any activities that could be perceived as offensive or hurtful by the local people, or that could harm the local environment. This moral imperative is nothing new: it was voiced by philosophers like Plato or Confucius thousands of years ago as a prerequisite for any travel.

For such encounters to be successful for both sides, those being visited must meet the visitors with hospitality, granting them the necessary protection and paying special attention to their vulnerability as outsiders. In particular, they should facilitate their access to all necessary information, safety, and assistance measures. Given the development of increasingly professional tourism, these requirements should largely be met, although there is still room for improvement—and is put to the test wherever mass tourism prevails.

This is illustrated by a little episode from Vienna, which might well have happened elsewhere.

Two policemen are standing in front of the Vienna Opera. A tourist approaches and asks one of them: "Do you speak English?" The policeman shakes his head. "Parlez-vous français?" The policeman shakes his head again. "Parlate italiano?" The same thing happens, and the tourist leaves in frustration. The other policeman says admiringly: "You should be able to speak so many languages!" to which the first replies: "So what—what good did it do him?"

Success as a Stress Factor

Tourism becomes a problem when it unreasonably alters the living environment of the local population. With regard to the phenomena of mass tourism, such as overtourism or overcrowding, this was already the case in many cities before the outbreak of the Covid pandemic, as well as in many holiday resorts on the Mediterranean or in the Alps. The city governments of Amsterdam and Barcelona have made major efforts to deal with the excesses of mass tourism. They try to encourage tourists to stay longer, seeing them as temporary residents of the city. They are trying to prevent tourist ghettos and a further concentration of existing tourist offers by means of visitor guidance measures. The aim is to improve the quality of life for local residents and tourists through better cooperation between the tourism industry, information centres and cultural institutions, as well as by adapting infrastructure and improving mobility.[4]

Sharing the city with visitors is a desired goal from the perspective of intercultural understanding, but it too requires regulation. Sharing the City is also a slogan of Airbnb. From the couch-surfing of young people interested in encountering other cultures, with hosts providing an air mattress or couch, Airbnb has become a highly profit-oriented industry. Its shareholders are the world's largest investment companies and bond fund providers, as well as global banks, none of which have anything in common with *Fairbnb*.

Valuable living space is lost for locals when owners rent their apartments to tourists solely for short-term profit increase. A harmonization is needed; and in Amsterdam, they have coined the term *Stad in Balans,* the balance of interests, integrating the various aspects into a development concept ultimately advantageous to all parties involved.

Measures include an 'Enjoy & Respect'-Campaign, increasing of the local tax to finance steering measures, the limitation of licenses for shops that offer nothing but tourist goods, the broadening of the attraction spectrum with partial re-direction of visitor flows (campaigns like 'See Amsterdam, Visit Holland ') and cooperation with various cultural institutions and museums, and encounter programmes. The restriction of car traffic and improvement of urban mobility — for example through a new subway line, priority for bicycles, etc.— and a large-scale house renovation programme *(Stadherstel Amsterdam)* indicate that the city practises all elements of a Governance Programme, an integrative urban development project which includes new technologies as well as moderated community processes.

The measures go far beyond tourist visitor management as such.[5] They were necessary to minimize the enormous burdens on the urban population. For example, in August 2017, more tourists slept in Amsterdam than did local residents—unusual for a city of millions, but normal for many Alpine villages or Mediterranean coastal destinations during the season. In the city centers of Amsterdam, Barcelona, Copenhagen, Lisbon, and Rome, each resident has tourists as neighbours for almost half of the year, or the population of the old town doubles during this time.

Forward-looking tourism policy measures—with destination management and new environmentally friendly transport solutions — could improve the scope for action as to how further tourism growth can be managed in the future. The increasing affordability of long-haul travel for Asians and the middle classes in emerging markets (in fact, air fares are far too low because aviation fuel is under-taxed), the rental of accommodation to city tourists, which puts pressure on the housing market, the reckless behaviour of thoughtless tourists—these are just some of the symptoms of *overtourism*. Added to this are the burdens of seasonal concentration, which further complicates the working conditions of service providers. The massive transport capacity of cruise ships and coaches brings the sudden arrival of large numbers of tourists. This overwhelms not only World Heritage cities such as Salzburg, Regensburg or Dubrovnik, but also other destinations. The dominance of these swarms of tourists can be countered by various forms of visitor control—but other factors of global tourism development can only be minimally influenced by destinations, if at all.[6]

Smart, Minimally Invasive Tourists

Avoiding such situations or coping with them is best achieved by flexible travellers who visit a destination well-informed. Philip Pearce[7] speaks of the *Smart Tourist*—the intelligent or experienced traveller who is aware of five principles before embarking on a journey. First, this tourist is well prepared for the destination and knows how to behave appropriately. Second, as a smart traveller, he or she takes advantage of the best mobility options, travelling by train or leaving the car on the outskirts of the city and using public transport. Such travellers can spend their time where there is something to see, without getting caught up in the dense traffic of big cities or fighting for parking spaces. They also book guided tours and use the discount cards available. As empathetic guests, they are respectful of the local people, behave properly and follow local conventions. This also makes it possible to get to know the local people as fellow human beings. Fourthly, such guests will devote sufficient time and attention to the place they visit, immersing themselves in the atmosphere in order to cause as little negative impact as possible. Fifthly, these tourists will also be smart technology users, using appropriate new technologies to optimise their stay.

These suggestions for smart tourism apply in principle to all travellers and all tourism destinations, but are particularly relevant for places under great pressure, like historical old towns or crowded winter sports areas. These all suffer from the same consequences of uncontrolled tourist flows, with ill-behaved guests leaving tons of garbage, producing hour-long traffic jams and creating enormous price-pressure on living space through second homes and short-term rentals.

There has been much talk of *Destination Marketing*—but in view of the likely growth in tourism, which needs to be developed is *Destination Management* involving all those affected, to correspond to a new form of cultural and urban development policy.[8] Such challenges also require political decisions, because appropriate laws and regulations are needed here. The problems become larger the longer active intervention is postponed and reliance is placed on 'self-regulation' of the markets. In the past, city governments usually contented themselves with leaving tourism development to the city marketing organizations. In focus was growth and attracting new guests, and records of overnight stays were published as success stories.

Overtourism or undertourism due to the pandemic and the climate crisis have redefined the measures and goals of future tourism. Tourism management must be seen in a much larger context, and requires more regulatory and political strategic planning intervention. It is not just a question of housing as many foreign guests as possible, but of taking into account the various dimensions of sustainability to the maximum extent.

Sustainable Tourism Management

From the experiences of the large European cultural cities in dealing with mass tourism, strategic approaches to sustainability can be seen from various perspectives.[9]

On the tourism service provider side, this primarily affects the bed supply: the restriction of further expansion of the tourism industry, the approval of new accommodations or second homes, and the licensing of Airbnb accommodations. Any further expansion should be subject to strict control or limitation. Variable pricing (seasonal dynamic pricing) can better align supply and demand. For ecological reasons alone, the accessibility of cities requires a reduction in car traffic and the enforced use of park & ride solutions, as well as the expansion of public mobility offers—which must of course also be available to local residents.

To control demand, communication services are needed, to attract the desired tourists or target groups and to keep others under control. This requires rules such as limiting the number of cruise ships that can dock at a port or the number of coaches that can approach a terminal at the same time. Here too, discounts or other benefits can be used to promote and strengthen the off-peak season or reduce peak loads. Tourism taxes can be used for regulation or optimization of visitor management. Information and rules of conduct are necessary to get conspicuous or intoxicated tourists to comply with local customary manners and conventions. The imposition of fines for public consumption of alcohol or urination outside of toilets, swimming in city fountains, and littering should serve as a warning, but also be spread through word of mouth. As the steps to the churches in Florence are regularly flooded with water, they remain clean—and the city administration prevents tourists from settling down for a snack and leaving their rubbish behind. This is also a creative solution to encourage visitors to the city to visit the many street cafés and restaurants: churches are not restaurants, but places of religious retreat and contemplation.

In some destinations, tourism limit caps will first be discussed and later introduced—perhaps not in terms of the number of tourists themselves, but of the number of licenses, permits for tourism infrastructure such as souvenir shops, hotel and private rooms/Airbnb, and parking space, as in Amsterdam and Dubrovnik. In this way, the sale of public space can also be countered—a goal that the Stadtverein (City Association) in Salzburg, founded in 1862 as the *Stadtverschönerungsverein* (City Beautification Association), is pursuing for the historic city, a UNESCO World Heritage Site. If the city squares are constantly played like a fair or always full of swarms of tourists, the special feature of the baroque ‚*Italian city*‘, no longer comes into its own, the World Heritage site loses its integrity as well as much of its special atmosphere.[10]

Back to Nature

One path that is already being taken leads to *nature-based tourism*. National park, biosphere park, alpine summer—these contexts not offer only the potential for economically profitable tourism, but also the possibility of stimulating regional economic value chains. *Salzburger*Land (the tourism brand name for the province of Salzburg) for example, with the highest density of organic farming in Europe, has the best prerequisites to succeed in this segment as a successful brand with a strong regional reference and without great 'staging' efforts. Tourism and leisure industry are a central field of action for regional policy. The mix of areas such as health, food, recreation, traditional healing methods, local crafts, and down-to-earth or nature-based lifestyle characterize the Salzburg region as a high-quality living and working space. Their careful valorization through quality tourism services can give both summer and winter tourism a unique signature. This is not just about monetary value creation and defining the region as a competitive unit: it also concerns the process of developing a vision for the living space, with the serious mediation of cultural heritage, the people, their ways of life, and their territorial identity.[11]

Nature-based tourism offers an excellent platform for experiencing the spectacles of nature—human nature included.[12] The direct encounter with the environment, local residents, their ways of life, their festivals and customs, the techniques with which they cultivate the land, preserve their flora and fauna, is decisive for the quality of experience. Immersion in this world, from which city-dwellers are too often alienated — indeed, the increasing everyday distance from nature, with negative health consequences, as described in the medical literature as *nature deficit disorder*[13] — can be countered through the magic of touch, through proximity and a process of the emotional appropriation of a space. Natural habitats stand as a counter-concept to urbanization, and are associated with human longings for peace, freedom, and harmony.

Due to their high scenic quality, many Alpine regions are highly coveted tourism destinations. Swiss researchers have developed indicators to determine quality standards for nature-based tourism.[14] To be classified as 'nature-based', a product or service must not cause any damage to the environment or landscape during the tourist's stay, it must minimise CO_2 emissions, and it must use soft mobility as much as possible. It is essential that providers are committed to protecting natural resources and regional characteristics, and that tourists have a certain ethical attitude as a motive for their visit. The protection of flora, fauna and landscape must be a priority in the design of tourism products. Nature-based tourism avoids the artificial production of experiences and noisy events: the focus is on experiencing nature as intensely as possible. Mottos such as "Let nature be nature" are not just slogans for the Schleswig–Holstein Wadden Sea National Park, the largest protected area between the North Cape and Sicily. They embody a vision for a different kind of tourism, one that is committed to respecting certain rules without neglecting the intense experience.[15]

This also sets a benchmark, distinguishing itself from areas where tourism, the entertainment industry, and the inevitable onslaught of masses of people have altered the character of the region, where no "honest" wine is served any longer, and the built-up infrastructure gives the landscape an urban look and feel.

Experiencing nature, rediscovering it, corresponds with the *evolutionistic worldview.* It marks the transition to smarter and sustainable technologies and moves the absolute primacy of humans away from centre-stage. The anthropic worldview of modernity, according to which *homo sapiens* could dominate everything with his technology, has long since collapsed. At the centre of the new thinking is the commonality of man and world or nature. It puts the world, and not its use for us, in the foreground.[16]

Credibility and sustainability are key criteria for this approach. If we can follow this vision and these quality criteria, the development of new ski areas, the further development of tourist infrastructure (chalet villages, cable cars, reservoirs, new roads, car parks, etc.) and other such functional uses of nature will no longer be an issue of conflict in local council meetings.

A regional development perspective towards sustainability also demands innovation towards *Soft or Gentle Mobility*.[17] Orienting towards Switzerland, which has developed its public transport in exemplary ways, would be a first step to reduce the ecological footprint of tourism. Since 1997, Werfenweng in *Salzburger*Land has been a model location for gentle mobility, with a forward-looking pilot project that advocates environmentally friendly travel.[18] Guests arrive by train, are met at the station, and electric vehicles are available. Having a private car is no longer necessary. Werfenweng is also a member of the *Alpine Pearls* network.[19] Altogether 28 municipalities from six nations have committed themselves to such responsible travel. As the first tourist cooperation, Alpine Pearls offers climate-neutral holidays.

A wide range of environmentally friendly offers focused on soft mobility contribute to climate protection. However, it must be acknowledged that CO_2 emissions cannot be completely avoided when people are on the move. Neukirchen am Großvenediger is also a member of this network, as are Hinterstoder in the Totes Gebirge mountain range and Mallnitz on the south side of the Hohe Tauern massif. The popular Bavarian tourist resorts Berchtesgaden and Bad Reichenhall are also Alpine Pearls, as is the World Natural Heritage site Vilnösser Tal in South Tyrol, as well as Arosa and Interlaken in Switzerland, and Bled in Slovenia. In 2011, the Alpine Pearls were awarded the Tourism for Tomorrow Award by the World Travel & Tourism Council.

A similar initiative was launched by the Austrian Alpine Club in 2008 with the *Bergsteigerdörfer* (Mountaineering Villages), which have been backed by the Alpine Clubs of neighbouring countries since 2016. They are supported by the Permanent Secretariat of the Alpine Convention to design tourism that is minimally invasive as regards nature and the cultural landscape. Strict criteria apply to the selection: location in the alpine area, the population must not exceed 2500, a down-to-earth tourism infrastructure, an alpine history, with mountain peaks as untouched as possible, well-managed mountain huts, clearly signposted network of paths, and local hiking and mountain guides. The focus is on a wide range of mountain sports and guided thematic hikes.

Conceptually, the aim is to offer guests holidays and nature-based recreation and to guarantee jobs for local residents in small catering businesses and mountain huts. Regionality is conveyed, for example, in the culinary arts and corresponds with the goal of sustainable economic development and the protection of mountain regions.

These Mountaineering Villages include Grünau in the Upper Austrian Almtal, Hüttschlag in the Salzburg Großarltal, Jezersko in the Slovenian Karawanks, Kreuth in the Bavarian Mangfall Mountains, Lungiarü in the South Tyrolean Dolomites, Ramsau in the Berchtesgaden Alps, the Sellraintal in the Stubai Alps, the Styrian Krakau in the Schladminger Tauern, Vent in the Ötztal and the East Tyrolean Villgratental.[20]

In terms of content, values are maintained, as is exemplified in the *Ecomuseum Simplon*. The concept is rooted in the idea of viewing the habitat holistically and demonstrating the interrelationship between humans and the environment. The Ecomuseum leaves the closed walls of museums and focuses on the natural space as well as the historical environment, which has transformed the natural landscape into a cultural landscape.[21] The main goals of such an Ecomuseum include serving as a 'database' for the local community and as an observation post for changes; being a focal point for gatherings, discussions, and innovations; and acting as a development laboratory for the community. Lastly, it serves visitors by enabling them to view the community and the region in a realistic showcase.[22]

The *Ecomuseum* project — developed in the 1970s by French museologists Georges-Henri Rivière and Hugues de Varine — is based on a holistic understanding of identity and the cultural heritage of a region. It also has a clear tourist reference and benefits from the trend of nature-based tourism, which highlights not only natural values but also the cultural dimension, combining both with considerations of sustainability and value creation. The intention is for guests to experience nature and culture authentically, and interact with the local population. This tourism should be developed with as little energy and interventions in the environment as possible and with the participation of the local people. The values and qualities of an area become visible and tangible for visitors, and closeness to nature serves as a guiding principle throughout the entire service chain — from environmentally friendly accommodation, to nutrition and to soft mobility.

Ten Commandments for the Alpine Pasture

Experiencing nature also requires competencies and rules of conduct, even very basic ones, as in the Australian Outback or Canadian wilderness. The same applies to the alpine pastures so popular with family vacationers. The tragic death of a German tourist on an alpine pasture in the Tyrolean Stubai Valley made headlines and led to a multi-year legal dispute. The woman, who was hiking with her dog, was attacked by a mother cow and fatally injured. The herd, consisting of ten approximately 700 kg cows and their ten calves, had probably become aggressive because of the dog. The autopsy revealed that the woman died from a crush injury to the pericardium. The herd owner was investigated for negligent homicide. In the first court case, he was sentenced, as the sole culprit, to a penalty of half a million euros. In the end, the Supreme Court ended the long legal dispute and decided on a verdict of partial guilt. With the amendment of the General Civil Code, the liability of livestock owners has recently been regulated. The new law provides for greater personal responsibility on the part of visitors to alpine pastures, in addition to the clear responsibility of the animal owners, who must follow rules of conduct on alpine pastures and meadows.[23]

As this was not an isolated case, and accidents on pastures kept happening, the traditional right of way has come up for discussion, and with it the restriction or even the end of alpine summer tourism. In the Salzburg region alone, there are more than 1800 managed alpine pastures; farmers traditionally cultivate the soils extensively with grazing animals and maintain the cultural landscape. Alpine pastures — supported by tourism associations and offers such as 'Pleasure hiking from pasture to pasture ' (as in the Großarl Valley, Salzburg Pongau)—have become true landscapes of longing for nature-loving people from the city. In the tourism media, nothing clouds the image of a picture-book landscape: the hardships of daily life on the fringes of the Alps are completely ignored. Visitors expect to find an alpine pasture where the dirndl-clad dairymaid pours the freshly tapped milk to accompany the cheese sandwich, and the softly mooing Pinzgau cattle contribute to the perfect postcard image that is sent via WhatsApp.

In fact, contradictions between alpine scenery and alpine reality do not exist, because the real activities on an alpine pasture remain almost invisible, and even the

10 The Vision: Smart Tourists, Minimally Invasive

animals enjoying their summer retreat there do not appear in the advertising materials. Perhaps visitors would be frightened when faced with a real alpine pasture populated with unpredictable fauna? Even the local alpine culture, the various work steps and activities until cheese can be served ready for consumption, or the local customs involved; these fail to find reflection in the advertising messages—there is only stylized idyll.[24]

This romantic touristic image of the alpine pasture fails to convey information on the serious problems that alpine farmers are struggling with. Too little public funding for their contribution to maintaining the alpine cultural landscape and the alpine personnel shortage, lamented for decades, have already led to a large extensification of alpine pastures. The income from tourist management often fails to cover the effort involved—and not all alpine farmers see themselves as 'experience hosts'. Remaining competitive with their hard-earned products and marketing them in self-distribution requires considerable additional efforts.[25] However, the alpine farmers have a fundamental interest in stocking and maintaining their alpine pastures, so it must therefore be a concern of all those involved to develop framework conditions for functioning combinations of alpine farming and tourism.

The 'Ten Commandments' or *committments,* rules of conduct for safe coexistence and respectful interaction with grazing livestock, have been developed as agreement between state governments and the responsible ministry.[26] They exemplify how the individual dream of freedom, often indulged in on vacation, reaches its limit and must fit into an order—for the benefit of all involved, humans and animals alike. Therefore, tourists are required to avoid contact with grazing livestock, not to feed the animals and to maintain safe distance from them; they are not to frighten the animals, must keep dogs under control and on a short leash. As mother cows protect their calves, encounters with dogs should be avoided. In general, the visitor should not leave the waymarked hiking trails on alpine pastures and meadows, and should pass grazing livestock at the greatest possible distance. All encounters with the people working here, nature, and the animals must be conducted with respect.

This does not require any sophisticated travel artistry—merely the observance of a few rules and some mindfulness. Controlling one's own behavior, adapting, and obeying any prohibitions, not becoming a factor of exploitation culturally and politically: this has always been central in developing country tourism and *tourism guidelines with insight*. This movement, supported by tourism-critical NGOs in the 1980s, was successful because several tourism operators gradually removed those of their products that were criticized as unethical, such as so-called 'human zoo trips' from their programmes. Moreover, the exploitation of children in tourism, forced into child labor or sexual services, has been largely stopped under the pressure of public campaigns.[27]

Sustainable Travel

In the wake of the Corona pandemic, vacationers and travellers were forced to adhere to various regulations and prohibitions, and it can be assumed that some of these restrictions will remain valid even after the end of the pandemic itself. This particularly concerns the protection of one's own health and the health of other people. Travellers adhere to this because they are directly affected—by the virus and the diseases involved, making them personally affected.

10 The Vision: Smart Tourists, Minimally Invasive

However, this subjective level of concern has not yet reached all population groups and tourism companies. With regard to the effects of climate change, the level of information and awareness has not yet led to a rethinking and to consistent climate-neutral action. In addition, the possible consequences of our current climate-damaging behavior lie in what must be seen as an uncertain future, although many phenomena are already evident. Lengthy heat waves, storm and flood disasters, winters without snow, glaciers without ice, and coastal areas without sandy beaches will massively change and affect tourism. It may well be that unbridled tourism, the all-inclusive of consumer frenzy, unrestrained travel mobility, and the principle of Anything Goes will lead to a dead end—although that end may not yet be in sight.

Travelling humanity is in a dilemma, because within us works a power-plant of irrationality that encourages contradiction. Not only philosophers, pondering important questions about God and the world, have been asking themselves whether we need to reinvent travel so that we can also cope better with staying at home.[28] However, philosophers generally do not think in prohibitions: their ethics rather corresponds to a culture of commandments. Therefore, we should recall Aristotle and his Middle Way between the two wrong ways, characterized by inadequacy and excess. This golden mean also plays a significant role in the teachings of His Holiness the 14th Dalai Lama, referring to an inner moderation, a self-restraint, equally distant from too much and too little. Here, mass tourism would be the exaggeration, the immoderation that comes from following exclusively hedonistic and economic interests. This could be contrasted by non-travel—which, however, does not go well with human curiosity, as global tourism has been seen as an "industry format of refined world curiosity of man".[29] The Middle Way—that would be the philosophical place to strive for, an ethical attitude involving: conscious living, use of the senses, consideration for nature, withdrawal of the ego. Careful use, cautious ways of dealing with our biodiversity should be seen as an absolute commandment. Participation in the world, which envisages one's own enjoyment, but ensures that not only one's own benefit is served—that would correspond to an ecological lifestyle that practices responsible action and an expanded self-understanding, which also builds a bridge to individuals in other lands, and future generations. This is not far from the *Humanistic travel ideal*, in which travel was seen as education of the heart, the mind, and the taste, and the miraculous did not necessarily grow in direct proportion to its distance. The world-renowned Asia correspondent Tiziano Terzani argued that one should at least set out on a journey with a question, and that such a journey is only meaningful if one returns with some answers in one's luggage.[30]

What is often unspecifically referred to as 'broadening of horizons' is a principle inherent in travel and vacationing, in every journey. Most people are driven by their curiosity, they want to grow beyond their own limitations and experience the world, which is why tourism always has a future. Tourists are individual treasure hunters like the 16th-century humanistic encyclopedist Theodor Zwinger, who in his *Methodus Academicus* speaks of the treasures of wisdom and virtue scattered over the world, to be collected and processed by clear thinking into instructions for daily life.[31]

The Austrian Emperor Joseph II, heir and successor to Empress Maria Theresa, had a very specific motive for travelling: to completely renew the Habsburg Empire and bring it into the modern age. An enlightened ruler, he travelled the entire monarchy between 1764 and 1787, seeking to understand the world in order to be able to change it. His curiosity and thirst for knowledge led him to remote regions where few foreigners had been, where there were neither roads nor maps. All trips were carefully planned in advance, but he travelled incognito, under a false name (but always in connection with the court in Vienna). This allowed him authentic encounters with state officials as well as with the common people. Of the 25 years of his reign, he spent about a quarter on the road—by carriage and on horseback. He travelled to see and hear everything, because he wanted to be able to draw his own conclusions, to learn from experience, in order to forge a reform programme. Indeed, Austria became a model country for state education policy: the social institutions created by him survived the centuries. As an enlightened thinker, he was far ahead of his time. In one of his letters he wrote, "If travel is useful for every thinking person, it is all the more so for a sovereign who, rejecting all pleasures, focuses only on the usefulness of his actions".[32]

To experience the world, to indulge in the pleasures of foreign places without becoming a burden there or elsewhere along the way—is this not the desirable goal of minimally invasive tourism?

Endnotes

1. Jean-Didier Urbain, *L'Idiot du voyage: Histoires de touristes*. Lausanne: Petite Bibliothèque Payot2002.
2. Martin Scharfe, *Berg-Sucht* (Mountain Addiction, A Cultural History of Early Alpinism), Vienna: Böhlau 2007.
3. https://www.unwto.org/ethics-culture-and-social-responsibility, 21.8.2021.
4. Greg Richards & Lenia Marques, *Creating synergies between cultural policy and tourism for permanent and*

temporary citizens. Committee on Culture of United Cities and Local Governments, Rotterdam 2018. http://www.agenda21culture.net, 4.3.2019.
5. Claartje van Ette, Amsterdam: A strategy to keep a growing city in balance, in: *Forum Mobil Extra—Salzburg Transport Days/Salzburg Tourism Forum* 2018, 18–20; on the governance approach see Alfred Kyrer & Bernhard Seyr (eds), *Governance und Wissensmanagement als wirtschaftliche Produktionsreserven* (Governance and Knowledge Management as Economic Productivity Reserves), Frankfurt: Lang 2007.
6. Harold Goodwin, Overtourism: Causes, Symptoms and Treatment, in: *Tourismus Wissen—Quarterly,* April 2019, 110–114.
7. Philip L. Pearce, *Limiting overtourism; the desirable new behaviours of the smart tourist.* http://tforum.today/2018/wp-content/uploads/2018/03/Limiting-overtourism-the-desirable-new-behaviours-of-the-smart-tourist.pdf; accessed 5.5.2019.
8. See Contributions in Kurt Luger & Matthias Ripp (Eds), *World Heritage, Place Making and Sustainable Tourism*, Innsbruck: StudienVerlag 2021.
9. Anja Saretzki & Karlheinz Wöhler (eds.), *Governance von Destinationen* (Governance of Destinations. New approaches for the successful control of tourist destinations), Berlin: Erich Schmidt 2013.
10. *Bastei—The Magazine of the Stadtverein* (City Association Salzburg). White Paper for the City of Salzburg, Spring 2019.
11. Kurt Luger, Nachhaltigkeitsüberlegungen zum Salzburg-Tourismus (Sustainability considerations for Salzburg tourism), in: Land Salzburg (ed.), *Weichenstellungen im Land Salzburg* (Setting the course in the Land Salzburg. Enquete of the state parliament, 9 October 2012). Salzburg: Publication series of the State Media Center 2012, 105–126.
12. Dominik Siegrist, Susanne Gessner & Lea Ketterer Bonnelame, *Naturaher Tourismus* (Nature-Based

10 The Vision: Smart Tourists, Minimally Invasive

Tourism, Quality Standards for Gentle Travel in the Alps), Berne: Haupt 2015.
13. Christina Pichler & Arnulf Hartl, Die Alpine Gesundheitsregion SalzburgerLand (The Alpine Health Region SalzburgerLand), in: Kurt Luger & Franz Rest (eds.), *Alpenreisen* (Alpine travels), Innsbruck: StudienVerlag 2017, 421–444.
14. https://www.naturnahertourismus.ch/index.php?id=14244, 23.8.2021.
15. https://www.nationalpark-wattenmeer.de/23.8.2021
16. Robert Pfaller & Klaus Kufeld (eds.), *Arkadien oder Dschungelcamp?* (Arcadia or Jungle Camp? Living in harmony or fighting with nature), Freiburg: Karl Alber 2014; Wolfgang Welsch, *Homo mundanus: Jenseits der anthropischen Denkform der Moderne* (Homo mundanus: Beyond the anthropic form of thought of modernity), Weilerswist: Velbrück 2012.
17. Christian Baumgartner, Nachhaltige Tourismus Entwicklung (Sustainable Tourism Development: Experiences from rural-alpine regions), in: Stephanie Brandl, Waldemar Berg, Marcus Herntrei, Christian Steckenbauer & Suzanne Lachmann-Falkner (eds.), *Tourismus und ländlicher Raum* (Tourism and rural space, Innovative strategies and tools for shaping the future), Berlin: Erich Schmidt 2021, 13–32.
18. www.werfenweng.eu
19. https://www.alpine-pearls.com
20. https://www.bergsteigerdoerfer.org/
21. https://www.ecomuseum.ch/;ecomuseums.com, 23.8.2021.
22. https://ecomuseums.com/ecomuseum-beginnings-hughes-de-varine-georges-henri-riviere-and-peter-davis/ 23.8.2021
23. https://www.agrarheute.com/land-leben/frau-kuh-getoetet-landwirt-490000-euro-zahlen-551930; https://www.krone.at/2152982, 25.8.2021.
24. Results of an empirical study by Maria Kirchner, *Kommunikation Alm* (Communication Alpine Pasture;

Native habitat and tourist place of longing), MA thesis, Salzburg 2017.
25. For details see Martin Anzengruber, *Almwirtschaft im Bundesland Salzburg* (Alpine Farming in the Federal State of Salzburg), Salzburg 2010.
26. https://www.Service.salzburg.gv.at/lkorrj/detail?nachrid=65466, Press release of the Salzburg State Correspondence, 13 August 2021.
27. https://www.ecpat.at/sexuelle-ausbeutung-von-kindern, 25.8.2021.
28. Klaus Kufeld wrote his valuable contribution about the future of travel in 2014, five years before the outbreak of the Covid pandemic. However, as far as climate change is concerned, the signs of the times were already stormy then. See his article: Vom Verlassen der Paradise (From Leaving Paradise), in: Roman Egger & Kurt Luger (eds), *Tourismus und mobile Freizeit* (Tourism and Mobile Leisure), Norderstedt: BoD 2015, 11–26.
29. Ibid., 14.
30. Tiziano Terzani, *Fliegen ohne Flügel* (Flying without Wings), Munich: Spiegel 1998.
31. Justin Stagl, *Eine Geschichte der Neugier* (A History of Curiosity), Vienna: Böhlau 2002, 158.
32. Monika Czernin, *Der Kaiser reist incognito* (The Emperor Travels Incognito, Joseph II and the Europe of Enlightenment), Munich: Penguin 2021.

GPSR Compliance
The European Union's (EU) General Product Safety Regulation (GPSR) is a set
of rules that requires consumer products to be safe and our obligations to
ensure this.

If you have any concerns about our products, you can contact us on

ProductSafety@springernature.com

In case Publisher is established outside the EU, the EU authorized
representative is:

Springer Nature Customer Service Center GmbH
Europaplatz 3
69115 Heidelberg, Germany

www.ingramcontent.com/pod-product-compliance
Lightning Source LLC
LaVergne TN
LVHW040734250326
834688LV00031B/290